Guide to MICROSCOPY

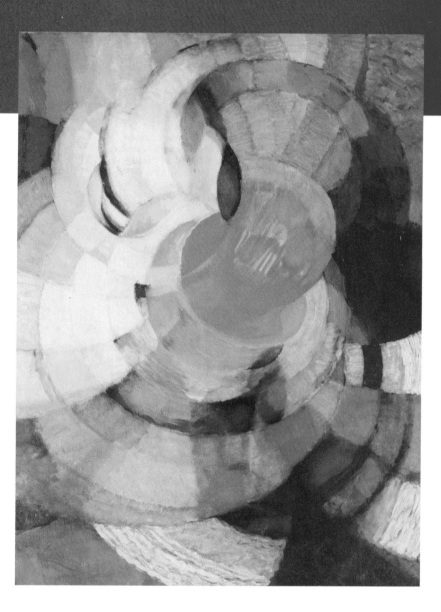

Wayne M. Becker
*University of Wisconsin,
Madison*

Lewis J. Kleinsmith
*University of Michigan,
Ann Arbor*

Jeff Hardin
*University of Wisconsin,
Madison*

San Francisco Boston New York
Capetown Hong Kong London Madrid Mexico City
Montreal Munich Paris Singapore Sydney Tokyo Toronto

Illustration and Photo Credits

Fig. 6 Courtesy of Leica, Inc. Deerfield, IL. **Fig. 8** M. I. Walker/Photo Researchers, Inc. **Figs. 10, 14** Tim Ryan. **Fig. 15** Strome, S., et al, "Spindle dynamics and the role of β-tubulin in early *Caenorhabditis elegans* embryos," from *Molecular Biology of the Cell* 12: 1751-64, Fig. 8. © 2001 Reprinted with permission by the American Society for Cell Biology. Image courtesy Susan Strome, Indiana University. **Fig. 16a, b** © Karl Garsha, Beckman Institute for Advanced Science and Technology, University of Illinois. **Fig. 18** Courtesy Nikon USA. **Fig. 19** From S.M. Potter, "Vital Imaging: Two Photons Are Better than One," *Current Biology* 6: 1595-98, ©1996. **Fig. 20** © Shelley Sazer, Baylor College of Medicine. Image courtesy Applied Precision, Inc. **Fig. 21a-d** E. D. Salmon, *Trends in Cell Biology* 5: 154-58, Fig. 3. **Fig. 23** From Sasha Koulish and Ruth G. Kleinfield, *Journal of Cell Biology* 23(1964): 39. Reproduced by copyright permission of The Rockefeller University Press. **Fig. 24a** Courtesy of J. L. Carson, Biological Education Consultants. **Figs. 24b, 25a** Carl Zeiss, Inc., Thornwood, NY 10594. **Fig. 26a** Courtesy of Don W. Fawcett, M.D., Harvard Medical School. **Fig. 26b** K. Tanaka, Osaka University, Osaka, Japan. **Fig. 27** Courtesy of H. Ris. **Fig. 28a** Carl Zeiss, Inc., Thornwood, NY 10594. **Fig. 29a, b** RMC, Inc. **Fig. 31** Courtesy of Dr. Michael F. Moody. **Fig. 32** Omikron/Photo Researchers, Inc. **Fig. 36** From L. Orci and A. Perrelet, *Freeze-Etch Histology,* Heidelberg: Springer-Verlag, 1975. **Fig. 37** Courtesy of H. Ris. **Fig. 40a** Reprinted with permission from *Nature* 171 (1953): 740; Copyright 1953 Macmillan Magazines Limited. **Fig. 40b** Richard Wagner, UCSF Graphics. **Table 1(1)** ©Biophoto Associates/Photo Researchers, Inc. **Table 1(2, 4, 5)** ©David M. Phillips/Visuals Unlimited. **Table 1(3)** ©Ed Reschke. **Table 1(6)** Courtesy of Noran Instruments.

Editorial Director:	Frank Ruggirello
Sponsoring Editor:	Michele Sordi
Project Editor/ Production Manager:	Laura Kenney Editorial Services
Project Coordinator:	Electronic Publishing Services Inc., N.Y.C.
Designer, Text and Cover:	Jennifer Dunn
Art Editor:	Kelly Murphy
Illustrations:	Precision Graphics
Composition/Page Makeup:	Electronic Publishing Services Inc., N.Y.C.
Indexing:	Shane-Armstrong Information Systems
Prepress Supervisor:	Vivian McDougal
Marketing Manager:	Josh Frost
Publishing Assistant:	Michael McArdle

On the cover: "Disks of Newton, Study for Fugue" by Frank Kupka, 1912, Philadelphia Museum of Art/CORBIS.

ISBN 0-8053-4869-7

Copyright © 2003 Pearson Education, Inc., publishing as Benjamin Cummings, 1301 Sansome St., San Francisco, CA 94111. All rights reserved. Manufactured in the United States of America. This publication is protected by Copyright and permission should be obtained from the publisher prior to any prohibited reproduction, storage in a retrieval system, or transmission in any form or by any means, electronic, mechanical, photocopying, recording, or likewise. To obtain permission(s) to use material from this work, please submit a written request to Pearson Education, Inc., Permissions Department, 1900 E. Lake Ave., Glenview, IL 60025. For information regarding permissions, call 847/486-2635.

Many of the designations used by manufacturers and sellers to distinguish their products are claimed as trademarks. Where those designations appear in this book, and the publisher was aware of a trademark claim, the designations have been printed in initial caps or all caps.

3 4 5 6 7 8 9 10—VH—05 04 03 02

Guide to Microscopy

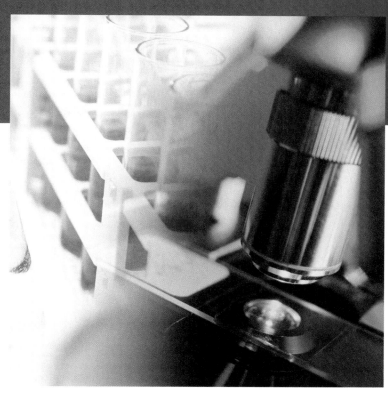

Cell biologists often need to examine the structure of cells and their components. The microscope is an indispensable tool for this purpose because most cellular structures are too small to be seen by the unaided eye. In fact, the beginnings of cell biology can be traced to the invention of the **light microscope,** which made it possible for scientists to see enlarged images of cells for the first time. The first generally useful light microscope was developed in 1590 by Z. Janssen and his nephew H. Janssen. Many important microscopic observations were reported during the next century, notably those of Robert Hooke, who observed the first cells, and Antonie van Leeuwenhoek, whose improved microscopes provided our first glimpses of internal cell structure. Since then, the light microscope has undergone numerous improvements and modifications, right up to the present time.

Just as the invention of the light microscope heralded a wave of scientific achievement by allowing us to see cells for the first time, the development of the **electron microscope** in the 1930s revolutionized our ability to explore cell structure and function. Because it is at least a hundred times better at visualizing objects than the light microscope, the electron microscope ushered in a new era in cell biology, opening our eyes to an exquisite subcellular architecture never before seen and changing forever the way we think about cells.

But in spite of its inferior resolving power, the light microscope has not fallen into disuse. To the contrary, light microscopy has experienced a renaissance in recent years as the development of specialized new techniques has allowed researchers to explore aspects of cell structure and behavior that cannot be readily studied by electron microscopy. These advances have involved the merging of technologies from physics, engineering, chemistry, and molecular biology, and they have greatly expanded our ability to study cells using the light microscope.

In this guide we will explore the fundamental principles of both light and electron microscopy, placing emphasis on the various specialized techniques that are used to adapt these two types of microscopy for a variety of specialized purposes.

Optical Principles of Microscopy

Although light and electron microscopes differ in many ways, they make use of similar optical principles to form images. Therefore we begin our discussion of microscopy by examining these underlying common principles, placing special emphasis on the factors that determine how small an object it is possible to see.

THE ILLUMINATING WAVELENGTH SETS A LIMIT ON HOW SMALL AN OBJECT CAN BE SEEN

Regardless of the kind of microscope being used, three elements are always needed to form an image: a *source of illumination,* a *specimen* to be examined, and a system of *lenses* that focuses the illumination on the specimen and forms the image. Figure 1 illustrates these features for a light microscope and an electron microscope. In a light microscope, the source of illumination is *visible light,* and the lens system consists of a series of glass lenses. The image can either be viewed directly through an eyepiece or focused on a detector, such as photographic film or an electronic camera. In an electron microscope, the illumination source is a *beam of*

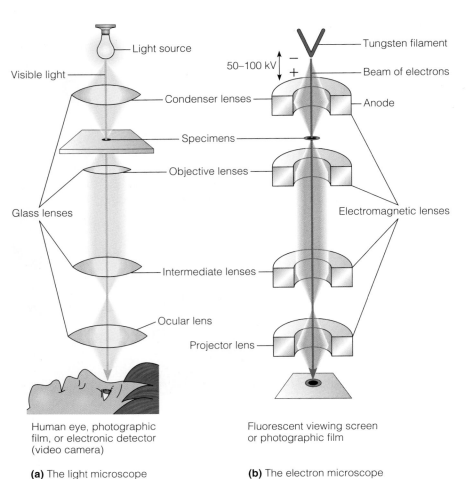

Figure 1 The Optical Systems of the Light Microscope and the Electron Microscope. **(a)** The light microscope uses visible light and glass lenses to form an image of the specimen that can be seen by the eye, focused on photographic film, or received by an electronic detector such as a video camera. **(b)** The electron microscope uses a beam of electrons emitted by a tungsten filament and focused by electromagnetic lenses to form an image of the specimen on a fluorescent screen or photographic film. (These diagrams have been drawn to emphasize the similarities in overall design between the two types of microscope. In reality, a light microscope is designed with the light source at the bottom and the ocular lens at the top, as shown in Figure 6b.)

electrons emitted by a heated tungsten filament, and the lens system consists of a series of electromagnets. The electron beam is focused either on a fluorescent screen for direct visualization of the image or on photographic film.

Despite these differences in illumination source and instrument design, both types of microscope depend on the same principles of optics and form images in a similar manner. When a specimen is placed in the path of a light or electron beam, physical characteristics of the beam are changed in a way that creates an image that can be interpreted by the human eye or recorded on a photographic detector. To understand this interaction between the illumination source and the specimen, we need to understand the concept of wavelength, which is illustrated in Figure 2 using the following simple analogy.

If two people hold onto opposite ends of a slack rope and wave the rope with a rhythmic up-and-down motion, they will generate a long, regular pattern of movement in the rope called a *wave form* (Figure 2a). The distance from the crest of one wave to the crest of the next is called the **wavelength.** If someone standing to one side of the rope tosses a large object such as a beach ball toward the rope, the ball may interfere with, or perturb, the wave form of the rope's motion (Figure 2b). However, if a small object such as a softball is tossed toward the rope, the movement of the rope will probably not be affected at all (Figure 2c). If the rope holders move the rope more rapidly, the motion of the rope will still have a wave form, but the wavelength will be shorter (Figure 2d). In this case, a softball tossed toward the rope is quite likely to perturb the rope's movement (Figure 2e).

This simple analogy illustrates an important principle: The ability of an object to perturb a wave motion depends crucially on the size of the object in relation to the wavelength of the motion. This principle is of great importance in microscopy, because it means that the wavelength of the illumination source sets a limit on how small an object can be seen. To understand this relationship, we need to recognize that the moving rope of Figure 2 is analogous to the beam of light (photons) or electrons that is used as an illumination source in a light or electron microscope, respectively—in other words, both light and electrons behave as waves. When a beam of light or electrons encounters a specimen, the specimen alters the physical characteristics of the illuminating beam, just as the beach ball or softball alters the motion of the rope. And because an object can be detected only by its effect on the wave, the wavelength must be comparable in size to the object that is to be detected.

Once we understand this relationship between wavelength and object size, we can readily appreciate why very small objects can be seen only by electron microscopy: The wavelengths of electrons are very much shorter than those of photons. Thus, objects such as viruses and ribosomes are too small to perturb a wave of photons, but they can readily interact with a wave of electrons. As we discuss different

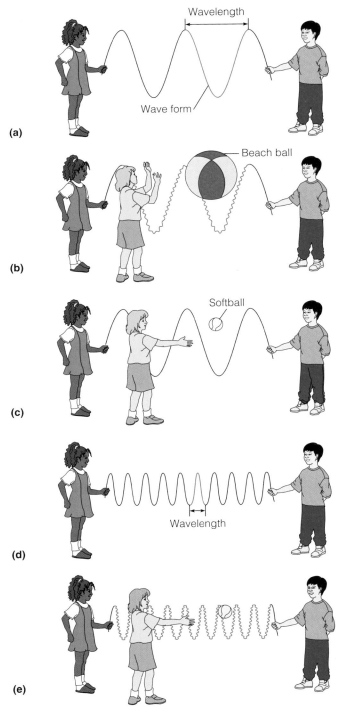

Figure 2 *Wave Motion, Wavelength, and Perturbations.* The wave motion of a rope held between two people is analogous to the wave form of both photons and electrons, and can be used to illustrate the effect of the size of an object on its ability to perturb wave motion. **(a)** Moving a slack rope up and down rhythmically will generate a wave form with a characteristic wavelength. **(b)** When thrown against a rope, a beach ball or other object with a diameter that is comparable to the wavelength of the rope will perturb the motion of the rope. **(c)** A softball or other object with a diameter significantly less than the wavelength of the rope will cause little or no perturbation of the rope. **(d)** If the rope is moved more rapidly, the wavelength will be reduced substantially. **(e)** A softball can now perturb the motion of the rope because its diameter is comparable to the wavelength of the rope.

types of microscopes and specimen preparation techniques, you might find it helpful to ask yourself how the source and specimen are interacting and how the characteristics of both are modified to produce an image.

RESOLUTION REFERS TO THE ABILITY TO DISTINGUISH ADJACENT OBJECTS AS SEPARATE FROM ONE ANOTHER

When waves of light or electrons pass through a lens and come to a focus, the image that is formed results from a property of waves called **interference**—the process by which two or more waves combine to reinforce or cancel one another, producing a wave equal to the sum of the two combining waves. Thus, the image that you see when you look at a specimen through a series of lenses is really just a pattern of either additive or canceling interference of the waves that went through the lenses, a phenomenon known as **diffraction.**

In a light microscope, glass lenses are used to direct the course of photons, whereas an electron microscope uses electromagnets as lenses to direct the course of electrons. Yet both kinds of lenses have two fundamental properties in common: focal length and angular aperture. The **focal length** is the distance between the midline of the lens and the point at which rays passing through the lens converge to a focus (Figure 3). The **angular aperture** is the half-angle α of the cone of light entering the objective lens of the microscope from the specimen (Figure 4). Angular aperture is therefore a measure of how much of the illumination that leaves the specimen actually passes through the lens. This in turn determines the sharpness of the interference pattern, and therefore the ability of the lens to convey information about the specimen. In the best light microscopes, the angular aperture is about 70°.

The angular aperture of a lens is one of the factors that influences a microscope's **resolution,** which is defined as the minimum distance that can separate two points that still remain identifiable as separate points when viewed through the microscope. To understand the role played by the angular aperture in determining resolution, we can ask how an infinitesimally small point of light is imaged by a simple lens. Figure 5 illustrates the type of pattern that is formed in

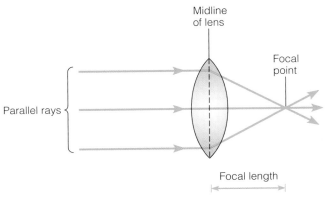

Figure 3 *The Focal Length of a Lens.* Focal length is the distance from the midline of a lens to the point at which parallel rays passing through the lens converge to a focus.

Optical Principles of Microscopy 3

the image plane under such conditions. Even though the original point of light (Figure 5a) is infinitesimal, the resulting image formed by the lens is not (Figure 5b and c). The image is a spot whose size and shape are the result of the diffraction of light going through the lens.

The size and shape of the spot that arises from an infinitesimal object will limit how close two small objects can be and still be resolved by the lens. Resolution, and hence the size of this *diffraction-limited spot,* is governed by three factors: the wavelength of the light used to illuminate the specimen, the angular aperture, and the refractive index of the medium surrounding the specimen. (**Refractive index** is a measure of the change in the velocity of light as it passes from one medium to another.) The effect of these three variables on resolution is described quantitatively by the following equation:

$$r = \frac{0.61\lambda}{n \sin \alpha} \quad (1)$$

where r is the resolution, λ is the wavelength of the light used for illumination, n is the refractive index of the medium between the specimen and the objective lens of the microscope, and α is the angular aperture as already defined. The constant 0.61 represents the degree to which image points can overlap and still be recognized as separate points by an observer.

In the preceding equation, the quantity $n \sin \alpha$ is called the **numerical aperture** of the objective lens, abbreviated **NA**. An alternative expression for resolution is therefore

$$r = \frac{0.61\lambda}{NA} \quad (2)$$

THE PRACTICAL LIMIT OF RESOLUTION IS ROUGHLY 200 NM FOR LIGHT MICROSCOPY AND 2 NM FOR ELECTRON MICROSCOPY

Maximizing resolution is an important goal in both light and electron microscopy. Because r is a measure of how close two points can be and still be distinguished from each other, resolution improves as r becomes smaller. Thus, for the best resolution, the numerator of equation 2 should be as small as possible and the denominator should be as large as possible.

We will begin by asking how to maximize resolution for a glass lens with visible light as the illumination source. First, we need to make the numerator as small as possible. The wavelength for visible light falls in the range of 400–700 nm, so the minimum value for λ is set by the shortest wavelength in this range that is practical to use for illumination, which turns out to be blue light of about 450 nm. To maximize the denominator of equation 2, recall that the numerical aperture is the product of the refractive index and the sine of the angular aperture. Both of these values must therefore be maximized to achieve optimal resolution. Since the angular

Figure 4 *The Angular Aperture of a Lens.* The angular aperture is the half-angle α of the cone of light entering the objective lens of the microscope from the specimen. (**a**) A low-aperture lens (α is small). (**b**) A high-aperture lens (α is large). The larger the angular aperture, the more information the lens can transmit. The best glass lenses have an angular aperture of about 70°.

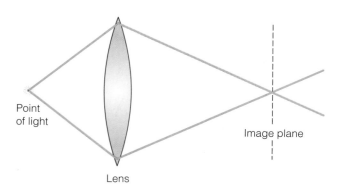
(a) Image formed from infinitesimal point of light

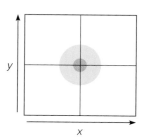
(b) Image created by exposing film in image plane

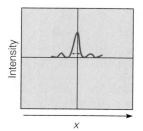
(c) Intensity plot of diffraction-limited spot in image plane

Figure 5 *Image Formation and the Resolution of a Lens.* (**a**) The image formed by a simple lens of an infinitesimally small point of light. In the plane where the rays come into focus, an interference pattern is formed. (**b**) The pattern formed on a piece of photographic film. (**c**) The trace of a scan of the film along the *x*-axis of part b. Even an infinitesimally small spot will have a finite size when imaged by a lens. The size of the spot determines how close two objects can be and still be resolved.

aperture for the best objective lenses is about 70°, the maximum value for sin α is about 0.94. The refractive index of air is about 1.0, so for a lens designed for use in air, the maximum numerical aperture is about 0.94.

Thus, for a lens with an angular aperture of 70°, the resolution in air for a sample illuminated with blue light of 450 nm can be calculated as follows:

$$r = \frac{0.61\lambda}{NA}$$
$$= \frac{(0.61)(450)}{0.94} = 292 \text{ nm} \qquad (3)$$

As a rule of thumb, then, the limit of resolution for a glass lens in air is roughly 300 nm.

As a means of increasing the numerical aperture, some microscope lenses are designed to be used with a layer of *immersion oil* between the lens and the specimen. Immersion oil has a higher refractive index than air and therefore allows the lens to receive more of the light transmitted through the specimen. Since the refractive index of immersion oil is about 1.5, the maximum numerical aperture for an oil immersion lens is about $1.5 \times 0.94 = 1.4$. The resolution of an oil immersion lens is therefore about 200 nm:

$$r = \frac{0.61\lambda}{NA}$$
$$= \frac{(0.61)(450)}{1.4} = 196 \text{ nm} \qquad (4)$$

Thus, the **limit of resolution** (best possible resolution) for a microscope that uses visible light is roughly 300 nm in air and 200 nm with an oil immersion lens. By using ultraviolet light as an illumination source, the resolution can be pushed to about 100 nm because of the shorter wavelength (200–300 nm) of this type of light. However, the image must then be recorded by photographic film or another detector because ultraviolet light is invisible to the human eye. Moreover, ordinary glass is opaque to ultraviolet light, so expensive quartz lenses must be used. In any case, these calculated values are theoretical limits to the resolution. In actual practice, such limits can rarely be reached because of aberrations (technical flaws) in the lenses.

Because the limit of resolution measures the ability of a lens to distinguish between two objects that are close together, it sets an upper boundary on the **useful magnification** that is possible with any given lens. In practice, the greatest useful magnification that can be achieved with a light microscope is about 1000 times the numerical aperture of the lens being used. Since numerical aperture ranges from about 1.0 to 1.4, this means that the useful magnification of a light microscope is limited to roughly 1000× in air and 1400× with immersion oil. Magnification greater than these limits is referred to as "empty magnification" because it provides no additional information about the object being studied.

The most effective way to achieve better magnification is to switch from visible light to electrons as the illumination source. Because the wavelength of an electron is about 100,000 times shorter than that of a photon of visible light, the theoretical limit of resolution of the electron microscope (0.002 nm) is orders of magnitude better than that of the light microscope (200 nm). However, practical problems in the design of the electromagnetic lenses used to focus the electron beam prevent the electron microscope from achieving this theoretical potential. The main problem is that electromagnets produce considerable distortion when the angular aperture is more that a few tenths of a degree. This tiny angle is several orders of magnitude less than that of a good glass lens (about 70°), giving the electron microscope a numerical aperture that is considerably smaller than that of the light microscope. The limit of resolution for the best electron microscope is therefore only about 0.2 nm, far from the theoretical limit of 0.002 nm. Moreover, when viewing biological samples, problems with specimen preparation and contrast are such that the practical limit of resolution is often closer to 2 nm. Practically speaking, therefore, resolution in an electron microscope is generally about 100 times better than that of the light microscope. As a result, the useful magnification of an electron microscope is about 100 times that of a light microscope, or about 100,000×.

The Light Microscope

It was the light microscope that first opened our eyes to the existence of cells. A pioneering name in the history of light microscopy is that of Antonie van Leeuwenhoek, the Dutch shopkeeper who is generally regarded as the father of light microscopy. Leeuwenhoek's lenses, which he manufactured himself during the late 1600s, were of surprisingly high quality for his time and were capable of 300-fold magnification, a tenfold improvement over previous instruments. This improved magnification made the interior of cells visible for the first time, and Leeuwenhoek's observations over a period of more than 25 years led to the discovery of cells in various types of biological specimens and set the stage for the formulation of the cell theory.

COMPOUND MICROSCOPES USE SEVERAL LENSES IN COMBINATION

In the 300 years since Leeuwenhoek's pioneering work, considerable advances in the construction and application of light microscopes have been made. Today, the instrument of choice for light microscopy uses several lenses in combination and is therefore called a **compound microscope** (Figure 6). The optical path through a compound microscope, illustrated in Figure 6b, begins with a source of illumination, usually a light source located in the base of the instrument. The light rays from the source first pass through **condenser lenses,** which direct the light toward a specimen mounted on a glass slide and positioned on the **stage** of the microscope. The **objective lens,** located immediately above the specimen, is responsible for forming the *primary image*. Most compound microscopes have several objective lenses of differing magnifications mounted on a rotatable turret.

The primary image is further enlarged by the **ocular lens,** or *eyepiece*. In some microscopes, an **intermediate lens** is positioned between the objective and ocular lenses to

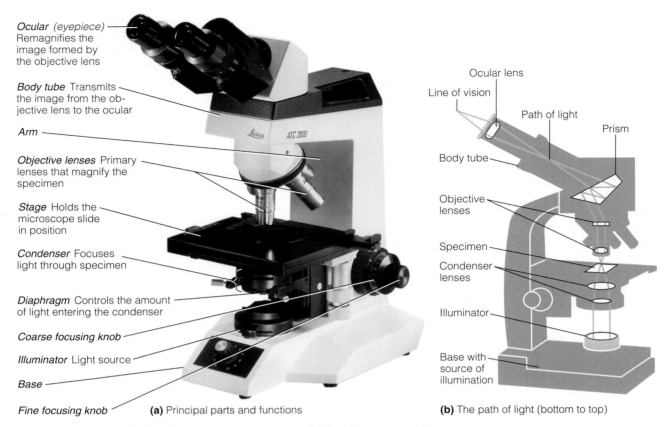

Figure 6 *The Compound Light Microscope.* **(a)** A compound light microscope. **(b)** The path of light through the compound microscope.

accomplish still further enlargement. You can calculate the overall magnification of the image by multiplying the enlarging powers of the objective lens, the ocular lens, and the intermediate lens (if present). Thus, a microscope with a 10× objective lens, a 2.5× intermediate lens, and a 10× ocular lens will magnify a specimen 250-fold.

The elements of the microscope described so far create a basic form of light microscopy called **brightfield microscopy.** Compared with other microscopes, the brightfield microscope is inexpensive and simple to align and use. However, the only specimens that can be seen directly by brightfield microscopy are those that possess color or have some other property that affects the amount of light that passes through. Many biological specimens lack these characteristics and must therefore be stained with dyes or examined with specialized types of light microscopes. Table 1 illustrates five such approaches by comparing human cheek epithelial cells as seen by brightfield microscopy (unstained and stained), fluorescence microscopy, phase-contrast microscopy, differential interference contrast microscopy, and confocal microscopy. We will look at these and several other important techniques in the following sections.

PHASE-CONTRAST MICROSCOPY DETECTS DIFFERENCES IN REFRACTIVE INDEX AND THICKNESS
As we will describe in more detail later, cells are often killed, sliced into thin sections, and stained before being examined by brightfield microscopy. While such procedures are useful for visualizing the details of a cell's internal architecture, little can be learned about the dynamic aspects of cell behavior by examining cells that have been killed, sliced, and stained. Therefore a variety of techniques have been developed for using light microscopy to observe cells that are intact and, in many cases, still living. One such technique, **phase-contrast microscopy,** improves contrast without sectioning and staining by exploiting differences in the thickness and refractive index of various regions of the cells being examined. To understand the basis of phase-contrast microscopy, we must first recognize that a beam of light is made up of many individual rays of light. As the rays pass from the light source through the specimen, their velocity may be affected by the physical properties of the specimen. Usually, the velocity of the rays is slowed down to varying extents by different regions of the specimen, resulting in a change in phase relative to light waves that have not passed through the object. (Light waves are said to be traveling *in phase* when the crests and troughs of the waves match each other.)

Although the human eye cannot detect such phase changes directly, the phase-contrast microscope overcomes this problem by converting phase differences into alterations in brightness. This conversion is accomplished using a *phase plate* (Figure 7), which is an optical material inserted into the light path above the objective lens to bring the direct or undiffracted rays into phase with those that have been dif-

Table 1 Different Types of Light Microscopy: A Comparison

Type of Microscopy	Light Micrographs of Human Cheek Epithelial Cells		Type of Microscopy
Brightfield (unstained specimen): Passes light directly through specimen; unless cell is naturally pigmented or artificially stained, image has little contrast.	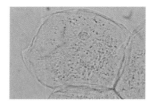	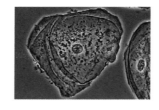	**Phase-contrast:** Enhances contrast in unstained cells by amplifying variations in refractive index within specimen; especially useful for examining living, unpigmented cells.
Brightfield (stained specimen): Staining with various dyes enhances contrast, but most staining procedures require that cells be fixed (preserved).	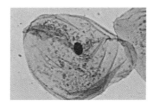		**Differential interference contrast:** Also uses optical modifications to exaggerate differences in refractive index.
Fluorescence: Shows the locations of specific molecules in the cell. Fluorescent substances absorb short wavelength radiation and emit longer wavelength, visible light. The fluorescing molecules may occur naturally in the specimen but more often are made by tagging the molecules of interest with fluorescent dyes or antibodies.	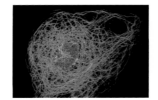	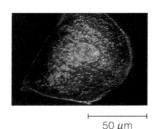 50 μm	**Confocal:** Uses lasers and special optics to focus illuminating beam on a single plane within the specimen. Only those regions within a narrow depth of focus are imaged. Regions above and below the selected plane of view appear black rather than blurry.

Source: From Campbell, Reece, and Mitchell, *Biology* 5th edition (Menlo Park, CA: Benjamin Cummings, 1999), p. 104.

fracted by the specimen. The resulting pattern of wavelengths intensifies the image, producing an image with highly contrasting bright and dark areas against an evenly illuminated background (Figure 8). As a result, internal structures of cells are often better visualized by phase-contrast microscopy than with brightfield optics.

This approach to light microscopy is particularly useful for examining living, unstained specimens because biological materials almost inevitably diffract light. Phase-contrast microscopy is widely used in microbiology and tissue culture research to detect bacteria, cellular organelles, and other small entities in living specimens.

Differential Interference Contrast (DIC) Microscopy Utilizes a Split Light Beam to Detect Phase Differences

Differential interference contrast (DIC) microscopy resembles phase-contrast microscopy in principle, but is more sensitive because it employs a special prism to split the illuminating light beam into two separate rays (Figure 9). When the two beams are recombined, any changes that occurred in the phase of one beam as it passed through the specimen cause it to interfere with the second beam. Because the largest phase changes usually occur at cell edges (the refractive index is more constant within the cell), the outline of the cell typically gives a strong signal. The image appears three-dimensional as a result of a shadow-casting illusion that arises because differences in phase are positive on one side of the cell but negative on the opposite side of the cell (Figure 10).

The optical components required for DIC microscopy consist of a *polarizer,* an *analyzer,* and a pair of *Wollaston prisms* (see Figure 9). The polarizer and the first Wollaston prism split a beam of light, creating two beams that are separated by a small distance along one direction. After traveling through the specimen, the beams are recombined by the second Wollaston prism. If no specimen is present, the beams recombine to form one beam that is identical to that which initially entered the polarizer and first Wollaston prism. In the presence of a specimen, the two beams do not recombine in the same way (i.e., they interfere with each other), and the resulting beam's polarization becomes rotated slightly compared with the original. The net effect is a remarkable enhancement in resolution that makes this technique especially useful for studying living, unstained specimens. As we will see shortly, combining this technique with video microscopy is an especially effective approach for studying dynamic events within cells as they take place.

Other contrast enhancement methods are also used by cell biologists. *Hoffman modulation contrast,* developed by Robert Hoffman, increases contrast by detecting optical gradients across a transparent specimen using special filters and a rotating polarizer. Hoffman modulation contrast results in a shadow-casting effect similar to DIC microspcopy.

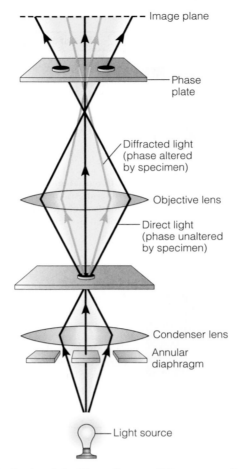

Figure 7 *Optics of the Phase-Contrast Microscope.* The configuration of the optical elements and the paths of light rays through the phase-contrast microscope. Pink lines represent light diffracted by the specimen, and black lines represent direct light.

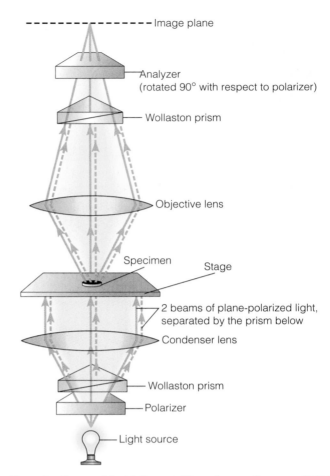

Figure 9 *Optics of the Differential Interference Contrast (DIC) Microscope.* The configuration of the optical elements and the paths of light rays through the DIC microscope.

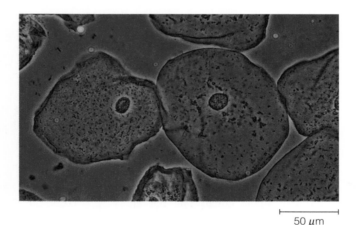

50 μm

Figure 8 *Phase-Contrast Microscopy.* A phase-contrast micrograph of epithelial cells. The cells were observed in an unprocessed and unstained state, which is a major advantage of phase-contrast microscopy.

FLUORESCENCE MICROSCOPY CAN DETECT THE PRESENCE OF SPECIFIC MOLECULES OR IONS WITHIN CELLS

Although the microscopic techniques described so far are quite effective for visualizing cell structures, they provide relatively little information concerning the location of specific molecules. One way of obtaining such information is through the use of **fluorescence microscopy,** which permits fluorescent molecules to be located within cells. To understand how fluorescence microscopy works, it is first necessary to understand the phenomenon of fluorescence.

The Nature of Fluorescence. The term **fluorescence** refers to a process that begins with the absorption of light by a molecule and ends with its emission. This phenomenon is best approached by considering the quantum behavior of light, as opposed to its wavelike behavior. Figure 11a is a diagram of the various energy levels of a simple atom. When an atom absorbs a photon (or *quantum*) of light of the proper energy, one of its electrons jumps from its ground state to a higher-energy, or *excited,* state. As the atom jiggles around, this electron often loses some of its energy and drops back down to the original ground state, emitting another photon as it does so. The emitted photon is always of less energy (longer wavelength) than the original photon that was absorbed. Thus, for example, shining blue light on the atom may result in red light being emitted. (The energy of a photon is inversely proportional to its wavelength; therefore, red light, being longer in wavelength than blue light, is lower in energy.)

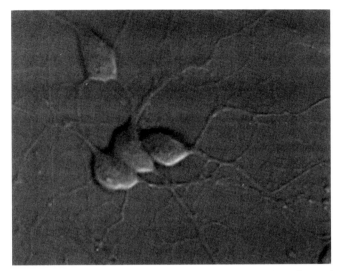

Figure 10 *DIC Microscopy.* A DIC micrograph of a cluster of rat hippocampal neurons growing in culture. Notice the shadow-casting illusion that makes these cells appear dark at the top and light at the bottom.

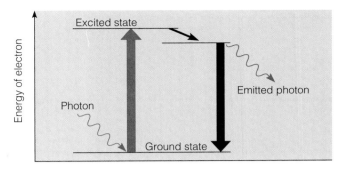

(a) Energy diagram

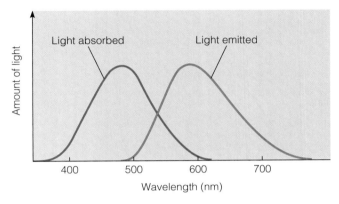

(b) Absorption and emission spectra

Figure 11 *Principles of Fluorescence.* **(a)** An energy diagram of fluorescence from a simple atom. Light of a certain energy is absorbed (e.g., the blue light shown here). The electron jumps from its ground state to an excited state. It returns to the ground state by emitting a photon of lower energy and hence longer wavelength (e.g., red light). **(b)** The absorption and emission spectra of a typical fluorescent molecule. The blue curve represents the amount of light absorbed as a function of wavelength, and the red curve shows the amount of emitted light as a function of wavelength.

Real fluorescent molecules have energy diagrams that are more complicated than that depicted in Figure 11a. The number of possible energy levels in real molecules is much greater, so the different energies that can be absorbed, and emitted, are correspondingly greater. The absorption and emission spectra of a typical fluorescent molecule are shown in Figure 11b. Every fluorescent molecule has its own characteristic absorption and emission spectra.

The Fluorescence Microscope. Fluorescence microscopy is a specialized type of light microscopy that employs light to excite fluorescence in the specimen. A fluorescence microscope has an *exciter filter* between the light source and the condenser lens that transmits only light of a particular wavelength (Figure 12). The condenser lens focuses the light on the specimen, causing fluorescent compounds in the specimen to emit light of longer wavelength. Both the excitation light from the illuminator and the emitted light generated by fluorescent compounds in the specimen then pass through the objective lens. As the light passes through the tube of the microscope above the objective lens, it encounters a *barrier filter* that specifically removes the excitation wavelengths. This leaves only the emission wavelengths to form the final fluorescent image, which therefore appears bright against a dark background.

Fluorescent Antibodies. To use fluorescence microscopy for locating specific molecules or ions within cells, researchers must employ special indicator molecules called *fluorescent probes*. A fluorescent probe is a molecule capable of emitting fluorescent light that can be used to indicate the presence of a specific molecule or ion.

One of the most common applications of fluorescent probes is in **immunofluorescence microscopy,** a technique based on the ability of antibodies to recognize and bind to specific molecules. (The molecules to which antibodies bind are called *antigens.*) Antibodies are proteins produced naturally by the immune system in response to an invading microorganism, but they can also be generated in the laboratory by injecting a foreign protein or other macromolecule into an animal such as a rabbit or mouse. In this way, it is possible to produce antibodies that will bind selectively to virtually any protein a scientist wishes to study. Antibodies are not directly visible using light microscopy, however, so they are linked to a fluorescent dye such as *fluorescein*, which emits a green fluorescence, or *rhodamine*, which emits a red fluorescence. To identify the subcellular location of a specific protein, cells are simply stained with a fluorescent antibody directed against that protein and the location of the fluorescence is then detected by viewing the cells with light of the appropriate wavelength.

Immunofluorescence microscopy can be performed using antibodies that are directly labeled with a fluorescent dye (Figure 13a). However, immunofluorescence microscopy is more commonly performed using **indirect immunofluorescence** (Figure 13b). In indirect immunofluorescence, a tissue or cell is treated

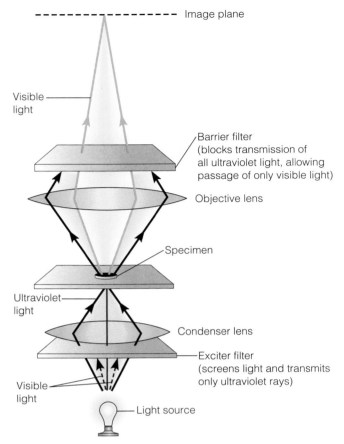

Figure 12 Optics of the Fluorescence Microscope. The configuration of the optical elements and the paths of light rays through the fluorescence microscope. Light from the source passes through an exciter filter that transmits only excitation light (solid black lines). Illumination of the specimen with this light induces fluorescent molecules in the specimen to emit longer-wavelength light (blue lines). The barrier filter subsequently removes the excitation light, while allowing passage of the emitted light. The image is therefore formed exclusively by light emitted by fluorescent molecules in the specimen.

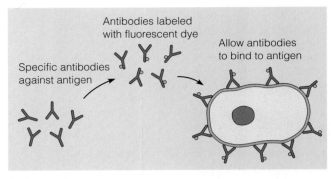

(a) Immunofluorescence

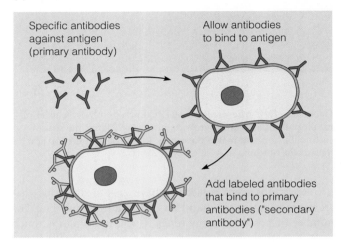

(b) Indirect immunofluorescence

Figure 13 Immunofluorescence Microscopy. Immunofluorescence microscopy relies on the use of fluorescently labeled antibodies to detect specific molecular components (antigens) within a tissue sample. **(a)** In direct immunofluorescence, an antibody that binds to a molecular component in a tissue sample is labeled with a fluorescent dye. The labeled antibody is then added to the tissue sample, and it binds to the tissue in specific locations. The pattern of fluorescence that results is visualized using fluorescence or confocal microscopy. **(b)** In indirect immunofluorescence, a primary antibody is added to the tissue. Then a secondary antibody that carries a fluorescent label is added. The secondary antibody binds to the primary antibody. Because more than one fluorescent secondary antibody can bind to each primary molecule, indirect immunofluorescence effectively amplifies the fluorescent signal, making it more sensitive than direct immunofluorescence.

with an antibody that is not labeled with dye. This antibody, called the *primary antibody*, attaches to specific antigenic sites within the tissue or cell. A second type of antibody, called the *secondary antibody*, is then added. The secondary antibody is labeled with a fluorescent dye, and it attaches to the primary antibody. Because more than one primary antibody molecule can attach to an antigen, and more than one secondary antibody molecule can attach to each primary antibody, more fluorescent molecules are concentrated near each molecule that we seek to detect. As a result, indirect immunofluorescence results in signal amplification, and it is much more sensitive than the use of a primary antibody alone. The method is "indirect" because it does not examine where antibodies are bound to antigens; technically, the fluorescence reflects where the secondary antibody is located. This, of course, provides an indirect measure of where the original molecule of interest is located.

Other Fluorescent Probes. Naturally occurring proteins that selectively bind to specific cell components are also used in fluorescence microscopy. For example, Figure 14 shows the fluorescence image of epithelial cells stained with a fluorescein-tagged mushroom toxin, *phalloidin,* which binds specifically to actin microfilaments. Another powerful fluorescence technique utilizes the *Green Fluorescent Protein (GFP),* a naturally fluorescent protein made by the jellyfish *Aequoria victoria.* Using recombinant DNA techniques, scientists can fuse DNA encoding GFP to a gene coding for a particular cellular protein. The resulting recombinant DNA can then be introduced into cells, where it is expressed to produce a fluorescent version of the normal cellular protein. In many cases, the fusion of GFP to the end of a protein does not interfere with its function, allowing the use of fluorescence microscopy to view the GFP-fusion protein as it functions in a living cell (Figure 15).

In addition to detecting macromolecules such as proteins, fluorescence microscopy can also be used to monitor the sub-

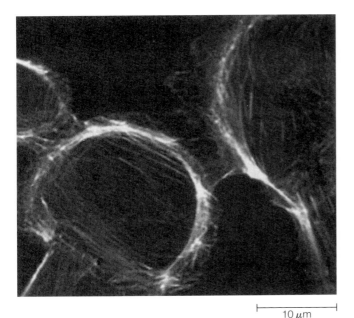

Figure 14 Fluorescence Microscopy. Cultured dog kidney epithelial cells stained with the fluorescent stain phalloidin, which binds to actin microfilaments.

cellular distribution of various ions. To accomplish this task, chemists have synthesized molecules whose fluorescent properties are sensitive to the concentrations of ions such as Ca^{2+}, H^+, Na^+, Zn^{2+}, and Mg^{2+}, as well as to the electrical potential across the plasma membrane. When these fluorescent probes are injected into cells in a form that becomes trapped in the cytosol or in a specific intracellular component, they provide important information about the ionic conditions inside the cell. For example, a fluorescent probe called *fura-2* is commonly used to track the Ca^{2+} concentration inside living cells because fura-2 emits a yellow fluorescence in the presence of low concentrations of Ca^{2+}, and a green and then blue fluorescence in the presence of progressively higher concentrations of this ion. Therefore, monitoring the color of the fluorescence in living cells stained with this probe allows scientists to observe changes in the intracellular Ca^{2+} concentration as they occur.

CONFOCAL MICROSCOPY MINIMIZES BLURRING BY EXCLUDING OUT-OF-FOCUS LIGHT FROM AN IMAGE

When intact cells are viewed, the resolution of fluorescence microscopy is limited by the fact that although fluorescence is emitted throughout the entire depth of the specimen, the viewer can focus the objective lens on only a single plane at any given time. As a result, light emitted from regions of the specimen above and below the focal plane cause a blurring of the image (Figure 16a). To overcome this problem, cell biologists often turn to the **confocal scanning microscope**—a specialized type of light microscope that employs a laser beam to produce an image of a single plane of the specimen at a time (Figure 16b). This approach improves the resolution along the optical axis of the microscope—that is, structures in the middle of a cell may be distinguished from those on the top or bottom. Likewise, a cell in the middle of a piece of tissue can be distinguished from cells above or below it.

To understand this type of microscopy, it is first necessary to consider the paths of light taken through a simple lens. Figure 17 illustrates how a simple lens forms an image of a point source of light. To understand what your eye would see, imagine placing a piece of photographic film in the plane of focus (image plane). Now ask how the images of other points of light placed further away or closer to the lens contribute to the original image (Figure 17b). As you might guess, there is a precise relationship between the distance of the object from the lens (o), the distance from the lens to the image of that object brought into focus (i), and the focal length of the lens (f). This relationship is given by the equation

$$\frac{1}{f} = \frac{1}{o} + \frac{1}{i} \qquad (5)$$

As Figure 17b shows, light arising from the points that are not in focus covers a greater surface area on the film because the rays are still either converging or diverging. Thus, the image on the film now has the original point source that is in focus, with a superimposed halo of light from the out-of-focus objects.

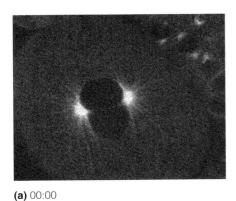
(a) 00:00

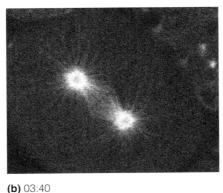
(b) 03:40

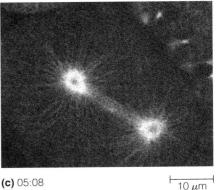
(c) 05:08

Figure 15 Using Green Fluorescent Protein to Visualize Proteins. An image series of a living, one-cell nematode worm embryo undergoing mitosis. The embryo is expressing β-tubulin that is tagged with the green fluorescent protein (GFP). Elapsed time from the first frame is shown in minutes : seconds.

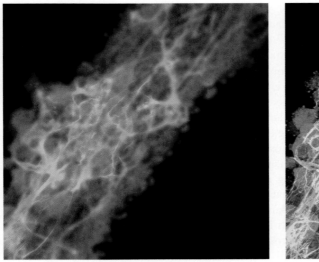 

(a) Traditional fluorescence microscopy   **(b)** Confocal fluorescence microscopy   25 μm

Figure 16 *Comparison of Confocal Fluorescence Microscopy with Traditional Fluorescence Microscopy.* These fluorescence micrographs show fluorescently labeled glial cells (red) and nerve cells (green) stained with two different fluorescent markers. **(a)** in traditional fluorescence microscopy the entire specimen is illuminated, so fluorescent material above and below the plane of focus tends to blur the image. **(b)** In confocal fluorescence microscopy, incoming light is focused on a single plane and out of focus fluorescence from the specimen is excluded. The resulting image is therefore much sharper. (Image Courtesy: Karl Garsha, Digital Light Microscopy Specialist, Imaging Technology Group, Beckman Institute for Advanced Science and Technology, University of Illinois at Urbana-Champaign, Urbana, Il, 61801 www.tig.uiux.edu)

If we were only interested in seeing the original point source, we could mask out the extraneous light by placing an aperture, or *pinhole,* in the same plane as the film. This principle is used in a confocal microscope to discriminate against out-of-focus rays. In a real specimen, of course, we do not have just a single extraneous source of light on each side of the object we wish to see, but a continuum of points. To understand how this affects our image, imagine that instead of three points of light, our specimen consists of a long thin tube of light, as in Figure 17c. Now consider obtaining an image of some arbitrary small section, dx. If the tube sends out the same amount of light per unit length, then even with a pinhole the image of interest will be obscured by the halos arising from other parts of the tube. This occurs because there is a small contribution from each out-of-focus section, and the sheer number of small sections will create a large background over the section of interest.

This situation is very close to that which we face when dealing with real biological samples that have been stained with a fluorescent probe. In general, the distribution of the probe is three-dimensional, and when we wish to look at the detail of a single object, such as a microtubule, using conventional fluorescence microscopy, the image is often marred by the halo of background light that arises mostly from microtubules above and below the plane of interest. To circumvent this, we can preferentially illuminate the plane of interest, thereby biasing the contributions in the image plane so that they arise mostly from a single plane (Figure 17d). Thus, the essence of confocal microscopy is to bring the illumination beam that excites the fluorescence into focus in a single plane, and to use a pinhole to ensure that the light we collect in the image plane arises mainly from that plane of focus.

Figure 18 illustrates how these principles are put to work in a laser scanning confocal microscope, which illuminates specimens using a laser beam focused by an objective lens down to a diffraction-limited spot (see Figure 5c). The position of the spot is controlled by scanning mirrors, which allow the beam to be swept over the specimen in a precise pattern. As the beam is scanned over the specimen, an image of the specimen is formed in the following way. First, the fluorescent light emitted by the specimen is collected by the objective lens and returned along the same path as the original incoming light. The path of the fluorescent light is then separated from the laser light using a *dichroic mirror,* which reflects one color but transmits another. Because the fluorescent light is lower in energy than the excitation beam, the fluorescence color is shifted toward the red. The fluorescent light passes through a pinhole placed at an image plane in front of a photomultiplier tube, which acts as a detector. The signal from the photomultiplier tube is then digitized and displayed by a computer. To see the enhanced resolution that results from confocal microscopy, look back to Figure 16, which shows images of the same cell visualized by conventional fluorescence microscopy and by laser scanning confocal microscopy.

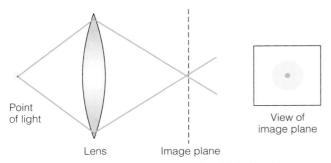

(a) Formation of an image of a single point of light by a lens

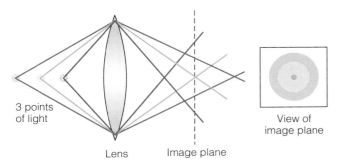

(b) Formation of an image of a point of light in the presence of two other points

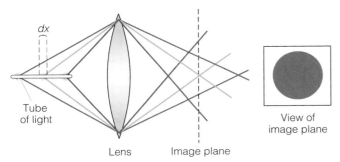

(c) Formation of an image of a section of an equally bright tube of light

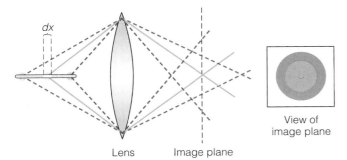

(d) Formation of an image of a brightened section of a tube of light

Figure 17 *Paths of Light Through a Single Lens.* **(a)** The image of a single point of light formed by a lens. **(b)** The paths of light from three points of light at different distances from the lens. In the image plane, the in-focus image of the central point is superimposed with the out-of-focus rays of the other points. A pinhole or aperture around the central point can be used to discriminate against out-of-focus rays and maximize the contributions from the central point. **(c)** The paths of light originating from a continuum of points, represented as a tube of light. This is similar to a uniformly illuminated sample. In the image plane, the contributions from an arbitrarily small in-focus section, *dx*, are completely obscured by the other out-of-focus rays; here a pinhole does not help. **(d)** By illuminating only a single section of the tube strongly and the rest weakly, we can recover information in the image plane about the section *dx*. Now a pinhole placed around the spot will reject out-of-focus rays. Because the rays in the middle are almost all from *dx*, we have a means of discriminating against the dimmer, out-of-focus points.

In confocal microscopy, a pinhole is used to exclude out-of-focus light. The result is a sharp image, but molecules above and below the focal plane of the objective lens are still being excited by the incoming light. This can result in rapid bleaching of the fluorescent molecules; in some cases, especially when viewing living cells that contain fluorescent molecules, such bleaching releases toxic radicals that can cause the cells to die. To reduce such "photodamage," it would be desirable if only the fluorescent molecules very close to the focal plane being examined were excited. This is possible using **multiphoton excitation microscopy**. In multiphoton excitation microscopy, a laser that emits pulses of light very rapidly and with very high energy is used to irradiate the specimen. Two (or in some cases three or more) photons must be absorbed by a fluorescent molecule in quick succession in order for it to fluoresce (Figure 19). The likelihood of this happening is very low, except near the focal plane of the objective lens. As a result, only the fluorescent molecules that are in focus fluoresce. The result is very similar in sharpness to confocal microscopy, but no pinhole is needed, as there is no out-of-focus light that needs to be excluded. Photodamage is also dramatically reduced. As an example, multiphoton excitation microscopy was used to image the living embryo shown in Figure 15.

A third technique, **digital deconvolution microscopy**, can be used to provide very sharp images. Digital deconvolution relies on a completely different principle. In this case, normal fluorescence microscopy is used to acquire a series of images throughout the thickness of a specimen. Then a computer is used to digitally process, or *deconvolve*, each focal plane to mathematically remove the contribution due to the out-of-focus light. In many cases, digital deconvolution can produce images that are comparable to those obtained by confocal microscopy (Figure 20). One advantage of deconvolution is that the microscope is not restricted to the specific wavelengths of light used in the lasers commonly found in confocal microscopes.

DIGITAL VIDEO MICROSCOPY CAN RECORD ENHANCED TIME-LAPSE IMAGES

The advent of solid-state light detectors has in many circumstances made it possible to replace photographic film with an electronic equivalent—that is, with a video camera or digital

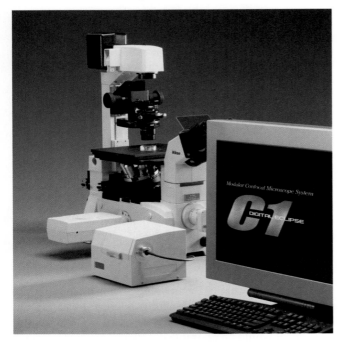

(a)

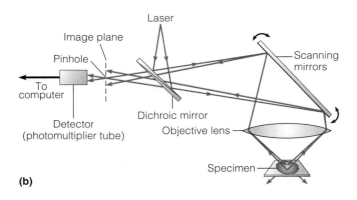

(b)

Figure 18 *A Laser Scanning Confocal Microscope.*
(a) A photograph and **(b)** a schematic of a laser scanning confocal microscope. A laser is used to illuminate one spot at a time in the specimen (blue lines). The scanning mirrors move the spot in a given plane of focus through a precise pattern. The fluorescent light that is emitted from the specimen (red lines) bounces off the same scanning mirrors and returns along the original path of the illumination beam. The emitted light does not return to the laser, but instead is transmitted through the dichroic mirror (which in this example reflects blue light but transmits red light). A pinhole in the image plane blocks the extraneous rays that are out of focus. The light is detected by a photomultiplier tube, the signal from which is digitized and stored by a computer.

imaging camera. These developments have given rise to the technique of **digital video microscopy,** in which microscopic images are recorded and stored electronically by placing a video camera in the image plane produced by the ocular lens. This approach takes advantage of the fact that video cameras can detect subtle differences in contrast better than the human eye. In addition, video cameras generate images in a digital form that can be *enhanced* by computer processing. In the enhancing process, the signal from the camera is first stored as a two-dimensional array of numbers whose values correspond to the lightness or darkness at particular locations in the image. The data are then processed by computer to increase contrast and remove background features that obscure the image of interest (Figure 21).

The resulting enhancement allows the visualization of structures that are an order of magnitude smaller than can be seen with a conventional light microscope. As we have seen, conventional light microscopy is generally incapable of resolving objects smaller than 200 nm in diameter. Computer-enhanced video microscopy, by contrast, permits the visualization of individual microtubules, which measure only 25 nm in diameter. Digital video techniques can be applied to conventional brightfield light microscopy, as well as to DIC and fluorescence microscopy, thereby creating a powerful set of approaches for improving the effectiveness of light microscopy.

An additional advantage of digital video microscopy is that the specimen does not need to be killed by fixation, as is required with electron microscopy, so dynamic events can be monitored as they take place. Moreover, special intensified video cameras have been developed that can detect extremely dim images, thereby facilitating the ability to record a rapid series of time-lapse pictures of cellular events as they proceed. For example, with conventional photographic film, it may take a minute or more to record the image of a fluorescently labeled cell, which means that a time-lapse series of photographs can only show one picture of the cell per minute. But an intensified video camera can record an image of the same cell 30 times per second, making it possible to monitor rapid changes in the appearance and behavior of subcellular components. This has allowed scientists to obtain information on the changes in concentration and subcellular distribution of such cytosolic components as second messengers during cellular signaling, and to study the role of cytoskeletal structures in intracellular movements. Thus, digital video microscopy has greatly expanded our ability to monitor events as they occur within living cells.

Digital microscopy is not only useful for examining events in one focal plane. In a variation of this technique, a computer is used to control a focus motor attached to a microscope. Images are then collected throughout the thickness of a specimen, When such a series of images is collected at specific time intervals, such microscopy is called *four-dimensional microscopy* (this phrase is borrowed from physics; the four dimensions are the three dimensions of space plus the additional dimension of time). Analyzing four-dimensional data requires special computer software that can navigate between focal planes over time to display specific images.

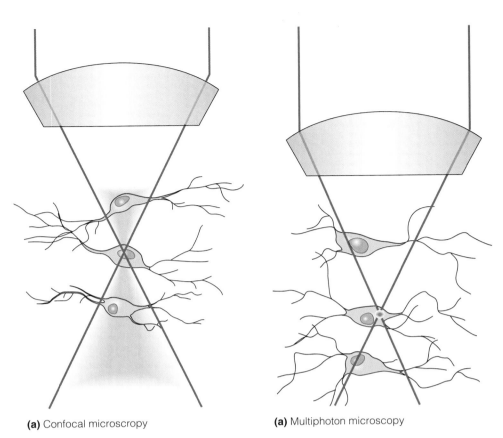

Figure 19 Multiphoton Excitation Microscopy. (a) In a standard laser scanning confocal microscope, the laser results in fluorescence in an hourglass-shaped path throughout the specimen. Because a large area fluoresces, photodamage is much more likely to occur than in multiphoton excitation microscopy. (b) In a multiphoton excitation microscope, fluorescence is limited to a spot at the focus of the pulsed infrared laser beam, resulting in much less damage. The infrared illumination also penetrates more deeply into the specimen than visible light.

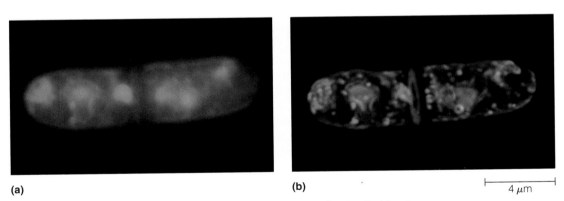

Figure 20 Digital Deconvolution Microscopy. A fission yeast cell stained with a dye specific for DNA (red) and a membrane-specific dye (green). The image on the left is an unprocessed optical section through the center of the cell. The image on the right is a projection of all the sections following three-dimensional image processing. The ring of the developing medial septum (red) is forming between the two nuclei (red) that arose by nuclear division during the previous mitosis.

Sample Preparation Techniques for Light Microscopy

One of the attractive features of light microscopy is the ease with which most specimens can be prepared for examination. In some cases, sample preparation involves nothing more than mounting a small piece of the specimen in a suitable liquid on a glass slide and covering it with a glass coverslip. The slide is then positioned on the specimen stage of the microscope and examined through the ocular lens, or with a camera.

However, to take maximum advantage of the resolving power of the light microscope, samples are usually prepared in a way designed to enhance *contrast*—that is, differences in the darkness or color of the structures being examined. A common means for enhancing contrast is to apply specific dyes that color or otherwise alter the light-transmitting properties of cell constituents. Preparing cells for staining typically involves special procedures, as we now describe.

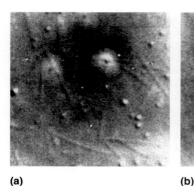 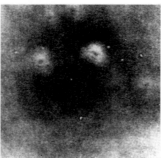 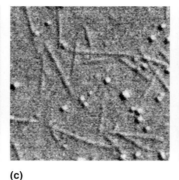 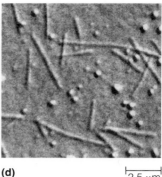

Figure 21 *Computer-Enhanced Digital Video Microscopy.* This series of micrographs shows how computers can be used to enhance images obtained with light microscopy. In this example, an image of several microtubules, which are too small to be seen with unenhanced light microscopy, are processed to make them visible in detail. **(a)** The image resulting from electronic contrast enhancement of the original image (which appeared to be empty). **(b)** The background of the enhanced image in (a), which is then **(c)** subtracted from image (a), leaving only the microtubules. **(d)** The final, detailed image resulting from electronic averaging of the separate images processed as shown in a–c.

Specimen Preparation Often Involves Fixation, Sectioning, and Staining

To prepare cells for staining, tissues are first treated with **fixatives** that kill the cells while preserving their structural appearance. The most widely employed fixatives are acids and aldehydes such as acetic acid, picric acid, formaldehyde, and glutaraldehyde. One way of fixing tissues is simply to immerse them in the fixative solution. An alternative approach for animal tissues is to inject the fixative into the bloodstream of the animal before removing the organs. This technique, called **perfusion,** may help reduce *artifacts,* which are false or inaccurate representations of the specimen that result from chemical treatment or handling of the cells or tissues.

In most cases, the next step is to slice the specimen into sections that are thin enough to transmit light. Otherwise, the specimen will appear opaque under the microscope and little discernible structure will be visible. To prepare such *thin sections,* the specimen is embedded in a medium that can hold it rigidly in position while sections are cut. The usual choice of embedding medium is paraffin wax. Since paraffin is insoluble in water, any water in the specimen must first be removed (by dehydration in alcohol, usually) and replaced by an organic solvent, such as xylene, in which paraffin is soluble. The processed tissue is then placed in warm, liquefied paraffin and allowed to harden. Dehydration is less critical if the specimen is embedded in a water-soluble medium instead of paraffin. Specimens may also be embedded in epoxy plastic resin, or as an alternative way of providing support, the tissue can simply be quick-frozen.

After embedding or quick-freezing, the specimen is sliced into thin sections a few micrometers thick by using a **microtome,** an instrument that operates somewhat like a meat slicer (Figure 22). The specimen is simply mounted on the arm of the microtome, which advances the specimen by small increments toward a metal or glass blade that slices the tissue into thin sections. As successive sections are cut, they usually adhere to one another, forming a ribbon of thin sections. These sections are then mounted on a glass slide and subjected to **staining** with any of a variety of dyes that have been adapted for this purpose. Sometimes the tissue is treated with a single stain, but more often a series of stains is used, each with an affinity for a different kind of cellular component. Once stained, the specimen is covered with a glass coverslip for protection.

Microscopic Autoradiography Locates Radioactive Molecules Inside Cells

Another approach for localizing specific components within cells is **microscopic autoradiography,** a technique that uses photographic emulsion to determine where a specific radioactive compound is located within a cell at the time the cell is fixed and sectioned for microscopy. In this procedure, shown in Figure 23, radioactive compounds are incubated with tissue sections or administered to intact cells or organisms. After sufficient time has elapsed for the radioactive compound to become incorporated into newly forming intracellular molecules and structures, the remaining unincorporated radioactivity is washed away and the specimen is sectioned in the conventional way and mounted on a microscope slide.

The slide is then covered with a thin layer of photographic emulsion and placed in a sealed box for the desired length of time, often for several days or even weeks. During this time, the radioactivity in the sample interacts with silver bromide crystals present in the photographic emulsion. When the emulsion is later developed and the specimen is examined under the microscope, *silver grains* appear directly above the specimen wherever radiation had bombarded the emulsion. The location of these silver grains, which are readily visible with both the light and electron microscope, can be used to pinpoint the region of the cell containing the radioactivity (Figure 24).

In practice, substances must emit relatively weak forms of radiation to be useful for microscopic autoradiography, because stronger radiation penetrates the emulsion too far to permit accurate localization. For this reason the most widely used radioisotope in autoradiography is tritium (^{3}H), an atom whose low-energy radiation permits a resolution of about 1 μm with the light microscope and close to 0.1 μm with the electron microscope. Since hydrogen is ubiquitous in biological molecules, a wide range of ^{3}H-labeled compounds are potentially available for use in autoradiography. For example, ^{3}H–amino acids are employed for locating newly synthesized proteins, deoxyribonucleosides such as ^{3}H-thymidine are used for monitoring DNA synthesis, ribonucleosides such as ^{3}H-uridine or ^{3}H-cytidine are employed for localizing newly made RNA molecules, and ^{3}H-glucose is used to study the synthesis of polysaccharides.

The Electron Microscope

The impact of electron microscopy on our understanding of cells can only be described as revolutionary. Yet, like light microscopy, electron microscopy has both strengths and weaknesses. In electron microscopy, resolution is much better, but specimen preparation and instrument operation are often more difficult. Electron microscopes are of two basic designs: the *transmission electron microscope* and the *scanning electron microscope*. Scanning and transmission electron microscopes are similar in that each employs a beam of electrons to pro-

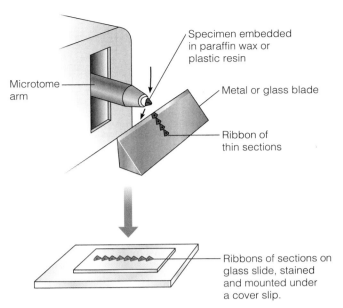

Figure 22 *Sectioning with a Microtome.* The fixed specimen is embedded in paraffin wax or plastic resin and mounted on the arm of the microtome. As the arm moves up and down through a circular arc, the blade cuts successive sections. The sections adhere to each other, forming a ribbon of thin sections that can be mounted on a glass slide, stained, and protected with a coverslip.

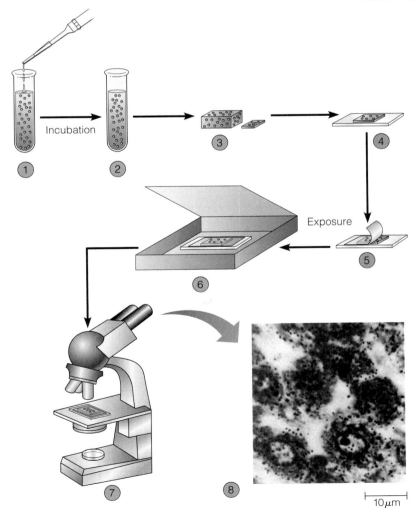

Figure 23 *Microscopic Autoradiography.* Microscopic autoradiography is a means of localizing radioactive molecules. ① To a suspension of cells, organisms, or tissue is added the desired radioactively labeled compound (blue solution) and the material is then incubated. ② The incubation is stopped, and the biological material is rinsed, fixed, and recovered from the fluid. ③ The fixed specimen is embedded and sectioned. ④ A section is washed and placed on a microscope slide. ⑤ The slide is covered with a thin layer of photographic emulsion (in a darkroom, of course) and then ⑥ placed in a sealed box for an appropriate length of time, to allow the radioactivity in the cell to expose the emulsion directly above it. ⑦ The slide is developed, rinsed, fixed, and washed. The specimen is then ready for examination under the microscope. ⑧ The inset is an autoradiograph of rat liver cells 12 hours after injection of the rat with ^{3}H-labeled cytidine, a precursor of RNA.

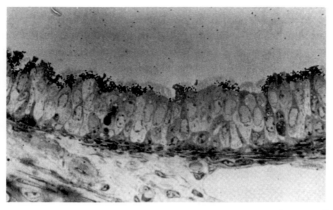 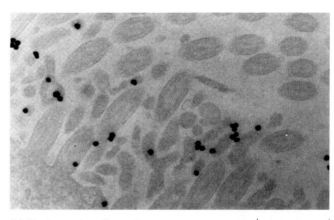

(a) Light autoradiograph $\quad$ 25 μm

(b) Electron autoradiograph $\quad$ 1 μm

Figure 24 *Autoradiographs Obtained by Light and Electron Microscopy.* Autoradiography can be applied to both light microscopy and electron microscopy. For the autoradiographs shown here, a radioactively labeled microbial pathogen was allowed to infect epithelial cells of the respiratory tract of an animal. The microbes can be identified by the black dots, which represent areas where the radioactive label exposed silver grains in the overlying photographic emulsion. **(a)** The limited resolving power of the light microscope precludes direct observation of the organisms, but their presence is indicated by the numerous black dots corresponding to exposed silver grains. **(b)** The higher magnification and resolution of the electron microscope reveal both the microbial cells and their identifying label (black dots).

duce an image. However, the instruments use quite different mechanisms to form the final image, as we see next.

TRANSMISSION ELECTRON MICROSCOPY FORMS AN IMAGE FROM ELECTRONS THAT PASS THROUGH THE SPECIMEN

The most commonly used type of electron microscope is called the **transmission electron microscope (TEM)** because it forms an image from electrons that are *transmitted* through the specimen being examined. As shown in Figure 25, most of the parts of the TEM are similar in name and function to their counterparts in the light microscope, although their physical orientation is reversed. We will look briefly at each of the major features.

The Vacuum System. Because electrons cannot travel very far in air, a strong vacuum must be maintained along the entire path of the electron beam. Two types of vacuum pumps work together to create this vacuum. On some TEMs, a device called a *cold trap* is incorporated into the vacuum system to help establish a high vacuum. The cold trap is a metal insert in the column of the microscope that is cooled by liquid nitrogen. The cold trap attracts gases and random contaminating molecules, which then solidify on the cold metal surface.

The Electron Gun. The electron beam in a TEM is generated by an **electron gun,** an assembly of several components. The *cathode,* a tungsten filament similar to a light bulb filament, emits electrons from its surface when it is heated. The cathode tip is near a circular opening in a metal cylinder. A negative voltage on the cylinder helps control electron emission and shape the beam. At the other end of the cylinder is the *anode.* The anode is kept at 0 V, while the cathode is maintained at 50–100 kV. This difference in voltage causes the electrons to accelerate as they pass through the cylinder and hence is called the **accelerating voltage.**

Electromagnetic Lenses and Image Formation. The formation of an image using electron microscopy depends on both the wavelike and the particle-like properties of electrons. Because electrons are negatively charged particles, their movement can be altered by magnetic forces. This means that the trajectory of an electron beam can be controlled using electromagnets, just as a glass lens can bend rays of light that pass through it.

As the electron beam leaves the electron gun, it enters a series of electromagnetic lenses (Figure 25b). Each lens is simply a space influenced by an electromagnetic field. The focal length of each lens can be increased or decreased by varying the amount of electric current applied to its energizing coils. Thus, when several lenses are arranged together, they can control illumination, focus, and magnification.

The **condenser lens** is the first lens to affect the electron beam. It functions in the same fashion as its counterpart in the light microscope to focus the beam on the specimen. Most electron microscopes actually use a condenser lens system with two lenses to achieve better focus of the electron beam. The next component, the **objective lens,** is the most important part of the electron microscope's sophisticated lens system. The specimen is positioned on the specimen stage within the objective lens. The objective lens, in concert with the **intermediate lens** and the **projector lens,** produces a final image on a *viewing screen* that fluoresces when struck by electrons.

How is an image formed from the action of these electromagnetic lenses on an electron beam? Recall that the electron beam generated by the cathode passes through the condenser lens system and impinges on the specimen. When the beam

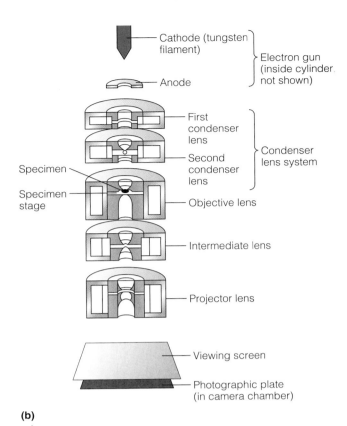

Figure 25 *A Transmission Electron Microscope.*
(a) A photograph and **(b)** a schematic diagram of a TEM.

strikes the specimen, some electrons are scattered by the sample, whereas others continue in their paths relatively unimpeded. This scattering of electrons is the result of properties created in the specimen by preparation procedures we will describe shortly. Specimen preparation, in other words, imparts selective *electron density* to the specimen; that is, some areas become more opaque to electrons than others. Electron-dense areas of the specimen will appear dark because few electrons pass through, whereas other areas will appear lighter because they permit the passage of more electrons.

The contrasting light, dark, and intermediate areas of the specimen create the final image. The fact that the image is formed by differing extents of electron transmission through the specimen is reflected in the name *transmission electron microscope*.

The Photographic System. Since electrons are not visible to the human eye, the final image is detected in the transmission electron microscope by allowing the transmitted electrons to strike a fluorescent screen or photographic film. The use of film allows one to create a photographic print called an **electron micrograph,** which then becomes a permanent photographic record of the specimen (Figure 26a). In some modern microscopes a digital camera records the screen or a digital detector directly detects incoming electrons.

Voltage. An electron beam is too weak to penetrate very far into biological samples, so specimens examined by conventional transmission electron microscopy must be extremely thin (usually no more than 100 nm in thickness). Otherwise, the electrons will not be able to pass through the specimen and the image will be entirely opaque. The examination of thicker sections requires a special **high-voltage electron microscope (HVEM),** which is similar to a transmission electron microscope but utilizes an accelerating voltage that is much higher—about 200–1000 kV compared with the 50–100 kV of a TEM. Because the penetrating power of the resulting electron beam is roughly ten times as great as that of conventional electron microscopes, relatively thick specimens can be examined with good resolution. As a result, cellular structure can be studied in sections as thick as 1 μm, about ten times the thickness possible with an ordinary TEM. Figure 27 shows a micrograph of a polytene chromosome from a fruit fly, as visualized using an HVEM.

SCANNING ELECTRON MICROSCOPY REVEALS THE SURFACE ARCHITECTURE OF CELLS AND ORGANELLES

Scanning electron microscopy is a fundamentally different type of electron microscopy that produces images from electrons deflected from a specimen's outer surface (rather than electrons transmitted through the specimen). It is an especially spectacular technique because of the sense of depth it gives to biological structures, thereby allowing surface topography to be studied (Figure 26b). As the name implies, a **scanning electron microscope (SEM)** generates such an image by scanning the specimen's surface with a beam of electrons.

An SEM and its optical system are shown in Figure 28. The vacuum system and electron source are similar to those found in the TEM, although the accelerating voltage is lower

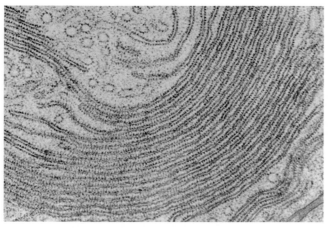
(a) Transmission electron micrograph ⊢0.5 μm⊣

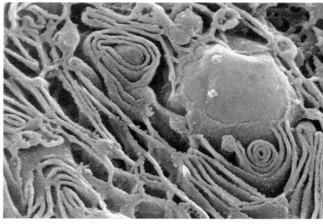
(b) Scanning electron micrograph ⊢1 μm⊣

Figure 26 *Comparison of Transmission and Scanning Electron Micrographs.* **(a)** The transmission electron micrograph shows membranes of rough endoplasmic reticulum in the cytoplasm of a rat pancreas cell. The "rough" appearance of the membranes in this specimen is caused by the presence of numerous membrane-bound ribosomes. **(b)** A similar specimen viewed by scanning electron microscopy reveals the three-dimensional appearance of the rough endoplasmic reticulum, although individual ribosomes cannot be resolved.

⊢0.5 μm⊣

Figure 27 *High-Voltage Electron Microscopy.* A polytene chromosome from the fruit fly *Drosophila melanogaster*, as seen with an HVEM operated at 1 million volts (1000 kV). For a three-dimensional view of this structure, see the stereo pair in Figure 37.

(about 5–30 kV). The main difference between the two kinds of instruments lies in the way the image is formed. In an SEM, the electromagnetic lens system focuses the beam of electrons into an intense spot that is moved back and forth over the specimen's surface by charged plates, called *beam deflectors*, that are located between the condenser lens and the specimen. The beam deflectors attract or repel the beam according to the signals sent to them by the deflector circuitry (Figure 28b).

As the electron beams sweeps rapidly over the specimen, molecules in the specimen are excited to high energy levels and emit *secondary electrons*. These emitted electrons are captured by a detector located immediately above and to one side of the specimen, thereby generating an image of the specimen's surface. The essential component of the detector is a *scintillator*, which emits photons of light when excited by electrons that impinge upon it. The photons are used to generate an electronic signal to a video screen. The image then develops point by point, line by line on the screen as the primary electron beam sweeps over the specimen.

Sample Preparation Techniques for Electron Microscopy

Specimens to be examined by electron microscopy can be prepared in several different ways, depending on the type of microscope and the kind of information the microscopist wants to obtain. In each case, however, the method is complicated, time-consuming, and costly compared with methods used for light microscopy. Moreover, living cells cannot be examined because of the vacuum to which specimens are subjected in the electron microscope.

ULTRATHIN SECTIONING AND STAINING ARE COMMON PREPARATION TECHNIQUES FOR TRANSMISSION ELECTRON MICROSCOPY

The most common way of preparing specimens for transmission electron microscopy involves slicing tissues and cells into ultrathin sections no more than 50–100 nm in thickness (less than one-tenth the thickness of the typical sections used for light microscopy). Specimens must first be chemically fixed and stabilized. The fixation step kills the cells but keeps the cellular components much as they were in the living cell. The fixatives employed are usually buffered solutions of aldehydes, most commonly glutaraldehyde. Following fixation, the specimen is often stained with a 1–2% solution of buffered osmium tetroxide (OsO_4), which binds to various components of the cell, making them more electron-dense.

The tissue is next passed through a series of alcohol solutions to dehydrate it, and then it is placed in a solvent such as acetone or propylene oxide to prepare it for embedding in liquefied plastic epoxy resin. After the plastic has infiltrated the specimen, it is put into a mold and heated in an oven to harden the plastic. The embedded specimen is then sliced into ultrathin sections by an instrument called an **ultramicrotome** (Figure 29a). The specimen is mounted firmly on the arm of the ultramicrotome, which advances the specimen in small increments toward a glass or diamond knife (Figure 29). When the block reaches the knife blade, ultrathin sections are cut from the block face. The sections float from the blade onto a water surface, where they can be picked up on a circular copper specimen grid. The grid consists of a meshwork of very thin copper strips, which support the specimen while still

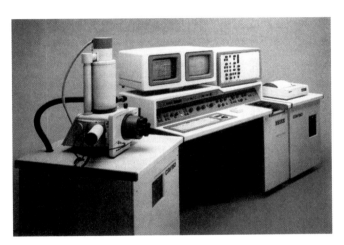

(a)

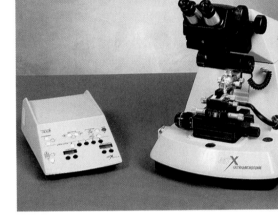

(a) Ultramicrotome

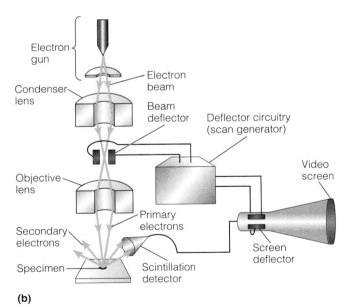

(b)

Figure 28 A Scanning Electron Microscope. (a) A photograph and (b) schematic diagram of an SEM. The image is generated by secondary electrons (short pink lines) emitted by the specimen as a focused beam of primary electrons (long pink lines) sweeps rapidly over it. The signal to the video screen is synchronized to the movement of the primary electron beam over the specimen by the deflector circuitry of the scan generator.

(b) Microtome arm of ultramicrotome

Figure 29 An Ultramicrotome. (a) A photograph of an ultramicrotome. (b) A close-up view of the ultramicrotome arm, showing the specimen in a plastic block mounted on the end of the arm. As the ultramicrotome arm moves up and down, the block is advanced in small increments, and ultrathin sections are cut from the block face by the diamond knife.

allowing openings between adjacent strips through which the specimen can be observed.

Once in place on the grid, the sections are usually stained again, this time with solutions containing lead and uranium. This step enhances the contrast of the specimen because the lead and uranium give still greater electron density to specific parts of the cell. After poststaining, the specimen is ready for viewing or photography with the TEM.

RADIOISOTOPES AND ANTIBODIES CAN LOCALIZE MOLECULES IN ELECTRON MICROGRAPHS

In our discussion of light microscopy, we described how microscopic autoradiography can be used to locate radioactive molecules inside cells. Autoradiography can also be applied to transmission electron microscopy, with only minor differences. For the TEM, the specimen containing the radioactively labeled compounds is simply examined in ultrathin sections on copper specimen grids instead of in thin sections on glass slides.

We also described how fluorescently labeled antibodies can be used in conjunction with light microscopy to locate specific cellular components. Antibodies are likewise used in the electron microscopic technique called **immunoelectron microscopy;** fluorescence cannot be seen in the electron microscope, so antibodies are instead visualized by linking them to substances that are electron dense and therefore visible as opaque dots. One of the most common approaches is to couple antibody molecules to colloidal gold particles. When ultrathin tissue sections are stained with gold-labeled antibodies directed against various proteins, electron microscopy can reveal the subcellular location of these proteins with great precision. (Figure 30).

NEGATIVE STAINING CAN HIGHLIGHT SMALL OBJECTS IN RELIEF AGAINST A STAINED BACKGROUND

Although cutting tissues into ultrathin sections is the most common way of preparing specimens for transmission electron microscopy, other techniques are suitable for particular purposes. For example, the shape and surface appearance of very small objects, such as viruses or isolated organelles, can be examined without cutting the specimen into sections. In the **negative staining** technique, which is one of the simplest techniques in transmission electron microscopy, intact specimens are simply visualized in relief against a darkly stained background.

To carry out negative staining, the copper specimen grid must first be overlaid with an ultrathin plastic film. The specimen is then suspended in a small drop of liquid, applied to the overlay, and allowed to dry in air. After the specimen has dried on the grid, a drop of stain such as uranyl acetate or phosphotungstic acid is applied to the film surface. The edges of the grid are then blotted in several places with a piece of filter paper to absorb the excess stain. This draws the stain down and around the specimen and its ultrastructural features. When viewed in the TEM, the specimen is seen in *negative contrast* because the background is dark and heavily stained, whereas the specimen itself is lightly stained (Figure 31).

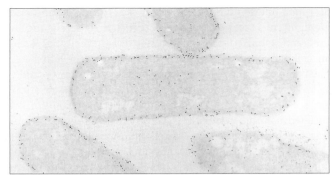

0.25 μm

Figure 30 *The Use of Gold-Labeled Antibodies in Electron Microscopy.* Cells of the bacterium *E. coli* were stained with gold-labeled antibodies directed against a plasma membrane protein. The small dark granules distributed around the periphery of the cell are the gold-labeled antibody molecules.

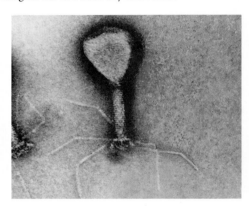

Figure 31 *Negative Staining.* An electron micrograph of a bacteriophage as seen in a negatively stained preparation. This specimen was simply suspended in an electron-dense stain, allowing it to be visualized in relief against a darkly stained background (TEM).

SHADOWING TECHNIQUES USE METAL VAPOR SPRAYED ACROSS A SPECIMEN'S SURFACE

Isolated particles or macromolecules can also be visualized by the technique of **shadowing** (Figure 32), which involves spraying a thin layer of an electron-dense metal such as gold or platinum at an angle across the surface of a biological specimen. Figure 33a illustrates the shadowing technique. The specimen is first spread on a clean mica surface and dried (step ①). It is then placed in a **vacuum evaporator,** a bell jar in which a vacuum is created by a system similar to that of an electron microscope (Figure 33b). Also within the evaporator are two electrodes, one consisting of a carbon rod located directly over the specimen and the other consisting of a metal wire positioned at an angle of about 10°–45° relative to the specimen.

After a vacuum is created in the evaporator, current is applied to the metal electrode, causing the metal to evaporate from the electrode and spray over the surface of the specimen (Figure 33a, ②). Because the metal-emitting electrode is positioned at an angle to the specimen, metal is deposited on only one side of the specimen, generating a metal *replica* of the surface. The opposite side of the specimen remains unstained; it is this unstained region that creates the "shadow" effect.

An overhead carbon-emitting electrode is then used to coat the specimen with evaporated carbon, thereby providing stability and support to the metal replicas (③). Next, the mica support containing the specimen is removed from the vacuum evaporator and lowered gently onto a water surface, causing the replica to float away from the mica surface. The replica is transferred into an acid bath, which dissolves away remaining bits of specimen, leaving a clean metal replica of the specimen (④). The replica is then transferred to a standard copper specimen grid (⑤) for viewing by transmission electron microscopy.

A related procedure is commonly used for visualizing purified molecules such as DNA and RNA. In this technique, a solution of DNA and/or RNA is spread on an air-water interface, creating a molecular monolayer that is collected on a thin film and visualized by uniformly depositing heavy metal on all sides.

FREEZE FRACTURING AND FREEZE ETCHING ARE USEFUL FOR EXAMINING THE INTERIOR OF MEMBRANES

Freeze fracturing is an approach to sample preparation that is fundamentally different from the methods described so far. Instead of cutting uniform slices through a tissue sample (or staining unsectioned material), specimens are rapidly frozen at the temperature of liquid nitrogen or liquid helium, placed in

Figure 32 Shadowing. An electron micrograph of tobacco mosaic virus particles visualized by shadowing. In this technique, heavy metal vapor was sprayed at an angle across the specimen, causing an accumulation of metal on one side of each virus particle and a shadow region lacking metal on the other side (TEM).

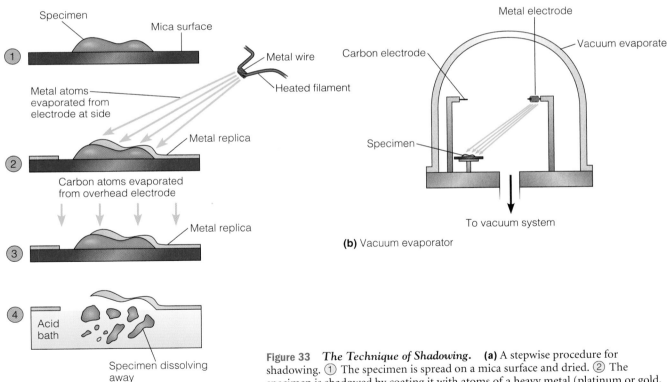

Figure 33 The Technique of Shadowing. (a) A stepwise procedure for shadowing. ① The specimen is spread on a mica surface and dried. ② The specimen is shadowed by coating it with atoms of a heavy metal (platinum or gold, shown in blue) that are evaporated from a heated filament located to the side of the specimen in a vacuum evaporator. This generates a metal replica (blue), the thickness of which reflects the surface contours of the specimen. ③ Next, the specimen is coated with carbon atoms evaporated from an overhead electrode to strengthen and stabilize the metal replica. ④ The replica is then floated onto the surface of an acid bath to dissolve away the specimen, leaving a clean metal replica. ⑤ The replica is washed and picked up on a copper grid for examination in the TEM. (b) The vacuum evaporator in which shadowing is done. The carbon electrode is located directly over the specimen, whereas the heavy metal electrode is off to the side.

a vacuum, and struck with a sharp knife edge. Samples frozen at such low temperatures are too hard to be cut. Instead, they fracture along lines of natural weakness—the hydrophobic interior of membranes, in most cases. Platinum/carbon shadowing is then used to create a replica of the fractured surface.

Freeze fracturing is illustrated in Figure 34. It takes place in a modified vacuum evaporator with an internal microtome knife for fracturing the frozen specimen. The temperature of the specimen support and the microtome arm and knife is precisely controlled. Specimens are generally fixed prior to freeze fracturing, although some living tissues can be frozen fast enough to keep them in almost lifelike condition. Because cells contain large amounts of water, fixed specimens are usually treated with an antifreeze such as glycerol to provide **cryoprotection**—that is, to reduce the formation of ice crystals during freezing.

The cryoprotected specimen is mounted on a metal specimen support (Figure 34, step ①) and immersed rapidly in freon cooled with liquid nitrogen (②). This procedure also reduces the formation of ice crystals in the cells. With the frozen specimen positioned on the specimen table in the vacuum evaporator (③), a high vacuum is established, the stage temperature is adjusted to around −100°C, and the frozen specimen is fractured with a blow from the microtome knife (④). A replica of the fractured specimen is made by shadowing with platinum and carbon as described in the previous section (⑤), and the replica is then ready to be viewed in the TEM (⑥).

Newcomers to the freeze-fracturing technique often misunderstand what a freeze-fracture replica represents. It is not the same as conventional sectioning with an ultramicrotome, which cuts through the specimen in a straight line (Figure 35a). Instead, the fracture line passes through the hydrophobic interior of membranes whenever possible, because this is the line of least resistance through the frozen specimen (Figure 35b). As a result, a freeze-fracture replica is largely a view of the interiors of membranes, showing the inside of one or the other of the two monolayers of the membrane.

Freeze-fractured membranes appear as smooth surfaces studded with **intramembranous particles (IMPs)** that are either randomly distributed in the membrane or organized into ordered complexes. These particles are integral membrane proteins that have remained with one lipid monolayer or the other as the fracture plane passes through the interior of the membrane.

The electron micrograph in Figure 36 shows the two faces of a plasma membrane revealed by freeze fracturing. The **P face** is the interior face of the inner monolayer; it is called the P face because this monolayer is on the *protoplasmic* side of the membrane. The **E face** is the interior face of the outer monolayer; it is called the E face because this monolayer is on the *exterior* side of the membrane. Notice that the P face has far more intramembranous particles than does the E face. In general, most of the particles in the membrane stay with the inner monolayer when the fracture plane passes down the middle of a membrane.

To have a P face and an E face appear side by side as in Figure 36, the fracture plane must pass through two neighboring cells, such that one cell has its cytoplasm and the inner monolayer of its plasma membrane removed to reveal the E face, while the other cell has the outer monolayer of its plasma membrane and the associated intercellular space

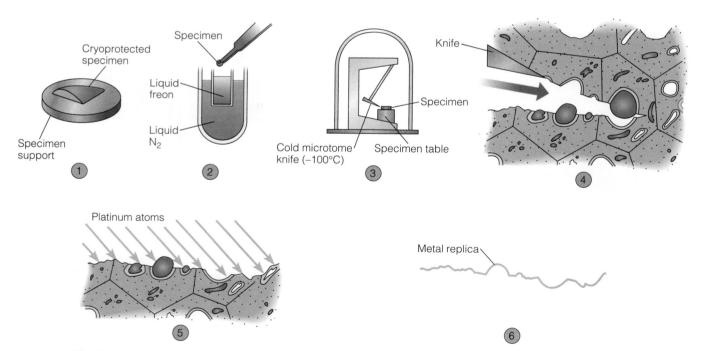

Figure 34 *The Technique of Freeze Fracturing.* ① A cryoprotected specimen is mounted on a metal support. ② The mounted specimen is immersed in liquid freon cooled in liquid nitrogen. ③ The frozen specimen is transferred to a vacuum evaporator and adjusted to a temperature of about −100°C. ④ The specimen is fractured with a blow from the microtome knife. The fracture plane passes through the interior of lipid bilayers wherever possible, because this is the line of least resistance through the frozen specimen, as shown in Figure 35b. ⑤ The fractured specimen is shadowed with platinum and carbon, as in Figure 33, to make a metal replica of the specimen. ⑥ The metal replica is examined in the TEM.

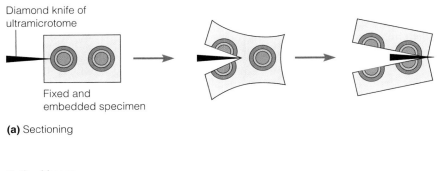
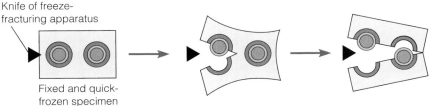

Figure 35 *Sectioning Versus Fracturing of Specimens.* **(a)** When a fixed and embedded specimen is sectioned for conventional transmission electron microscopy, the edge of the diamond knife makes a clean, linear cut through the tissue. **(b)** When a fixed and quick-frozen specimen is fractured, the blow of the knife generates a fracture plane through the frozen sample that passes through the interiors of membranes whenever possible, because the hydrophobic interior of a phospholipid bilayer is more readily fractured than is the ice that surrounds it. The interiors of membranes are therefore exposed on the fracture surfaces, as Figure 36 shows.

removed to reveal the P face. Accordingly, E faces are always separated from P faces of adjacent cells by a "step" (marked by the arrows in Figure 36) that represents the thickness of the intercellular space.

In a closely related technique called **freeze etching,** a further step is added to the conventional freeze-fracture procedure to make the technique even more informative. Following the fracture of the specimen but prior to shadowing, the microtome arm is placed directly over the specimen for a short period of time (a few seconds to several minutes). This maneuver causes a small amount of water to evaporate (sublime) from the surface of the specimen to the cold knife surface, and this sublimation produces an *etching* effect—that is, an accentuation of surface detail. Brief etching enhances the view of outer membrane surfaces, which are otherwise covered with ice and difficult to see in a typical freeze-fracture specimen.

By using ultrarapid freezing techniques to minimize the formation of ice crystals during freezing, and by including a volatile cryoprotectant such as aqueous methanol, which sublimes very readily to a cold surface, the etching period can be extended and a deeper layer of ice can be removed, thereby exposing structures that are located deep within the cell interior. This modification, called **deep etching,** provides a fascinating look at cellular structure. Deep etching has been especially useful in exploring the cytoskeleton and examining its connections with other structures of the cell.

STEREO ELECTRON MICROSCOPY ALLOWS SPECIMENS TO BE VIEWED IN THREE DIMENSIONS

Electron microscopists frequently want to visualize specimens in three dimensions. Shadowing, freeze fracturing, and scanning electron microscopy are useful for this purpose, as is another specialized technique called **stereo electron microscopy.** In stereo electron microscopy, three-dimensional information is obtained by photographing the same specimen at two slightly different angles. This is accomplished using a special specimen stage that can be tilted relative to the electron beam. The specimen is first tilted in one direction and photographed, then tilted an equal amount in the opposite direction and photographed again.

Figure 36 *Freeze Fracturing of the Plasma Membrane.* This electron micrograph shows the exposed faces of the plasma membranes of two adjacent endocrine cells from a rat pancreas as revealed by freeze fracturing. The P face is the inner surface of the lipid monolayer on the protoplasmic side of the plasma membrane. The E face is the inner surface of the lipid monolayer on the exterior side of the plasma membrane. The P face is much more richly studded with intramembranous particles than the E face. The arrows indicate the "step" along which the fracture plane passed from the interior of the plasma membrane of one cell to the interior of the plasma membrane of a neighboring cell. The step therefore represents the thickness of the intercellular space (TEM).

The two micrographs are then mounted side by side as a *stereo pair*. When you view a stereo pair through a stereoscopic viewer, your brain uses the two independent images to construct a three-dimensional view that gives a striking sense of depth to the structure under investigation. Figure 37 is a stereo pair of the *Drosophila* polytene chromosome seen earlier in a high-voltage electron micrograph (see Figure 27). Using a stereo viewer or allowing your eyes to fuse the two images visually creates a striking, three-dimensional view of the chromosome.

Specimen Preparation for Scanning Electron Microscopy Involves Fixation but Not Sectioning

When preparing a specimen for scanning electron microscopy, the goal is to preserve the structural features of the cell surface and to treat the tissue in a way that minimizes damage by the electron beam. The procedure is actually similar to the preparation of ultrathin sections for transmission electron microscopy, but without the sectioning step. The tissue is fixed in aldehyde, postfixed in osmium tetroxide, and dehydrated by processing through a series of alcohol solutions. The tissue is then placed in a fluid such as liquid carbon dioxide in a heavy metal canister called a **critical point dryer,** which is used to dry the specimen under conditions of controlled temperature and pressure. This helps keep structures on the surfaces of the tissue in almost the same condition as they were before dehydration.

The dried specimen is then attached to a metal specimen mount with a metallic paste. The mounted specimen is coated with a layer of gold or a mixture of gold and palladium, using a modified form of vacuum evaporation called **sputter coating.** Once the specimen has been mounted and coated, it is ready to be examined in the SEM.

Other Imaging Methods

Light and electron microscopy are direct imaging techniques in that they use photons or electrons to produce actual images of a specimen. However, some imaging techniques are indirect. To understand what we mean by indirect imaging, suppose you are given an object to handle with your eyes closed. You might feel 6 flat surfaces, 12 edges and 8 corners, and if you then draw what you have felt, it would turn out to be a box. This is an example of an indirect imaging procedure.

The two indirect imaging methods we describe here are *scanning probe microscopy* and *X-ray diffraction*. Both approaches have the potential for showing molecular structures at near-atomic resolution, ten times better than the best electron microscope. They do have some shortcomings that limit their usefulness with biological specimens, but when these techniques can be applied successfully, the resulting images provide unique information about molecular structure that cannot be obtained using conventional microscopic techniques.

Scanning Probe Microscopy Reveals the Surface Features of Individual Molecules

Although "scanning" is involved in both scanning electron microscopy and scanning probe microscopy, the two methods are in fact quite different. The first example of a **scanning probe microscope,** called the *scanning tunneling microscope (STM),* was developed in the early 1980s for the purpose of exploring the surface structure of specimens at the atomic level. The STM utilizes a tiny probe that does not emit an electron beam, but instead possesses a tip made of a conducting material such as platinum-iridium. The tip of the probe is extremely sharp; ideally its point is composed of a single atom. It is under the precise control of an electronic circuit that can move it in three dimensions over a surface. The x and y dimensions scan the surface, while the z dimension governs the distance of the tip above the surface (Figure 38).

As the tip of the STM is moved across the surface of a specimen, voltages from a few millivolts to several volts are applied. If the tip is close enough to the surface and the surface is electrically conductive, electrons will begin to leak or "tunnel" across the gap between the probe and the sample. The tunneling is highly dependent on the distance, so that even small irregularities in the size range of single atoms will affect the rate of electron tunneling. As the probe scans the sample, the tip of the probe is automatically moved up and

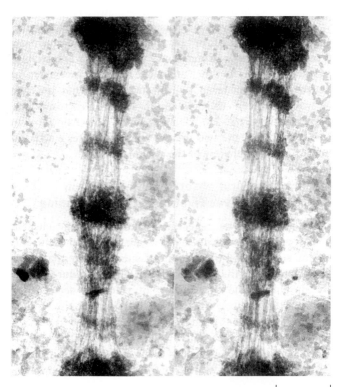

Figure 37 *Stereo Electron Microscopy.* The polytene chromosome seen in the high-voltage electron micrograph of Figure 27 is shown here as a stereo pair of photos that can be fused optically to generate a three-dimensional image. The two photographs were taken by tilting the specimen stage first 5° to the right, then 5° to the left of the electron beam. For a three-dimensional view, a stereo pair can be examined with a stereoscopic viewer. Alternatively, simply let your eyes cross slightly, fusing the two micrographs into a single image (TEM).

down to maintain a constant rate of electron tunneling across the gap. A computer measures this movement and uses the information to generate a map of the sample's surface, which is viewed on a video screen.

In spite of the enormous power of the STM, it suffers from two limitations: the specimen must be an electrical conductor, and the technique only provides information about electrons associated with the specimen's surface. Researchers have therefore begun to develop other kinds of scanning probe microscopes that scan a sample just like the STM, but measure different kinds of interactions between the tip and the sample surface. For example, in the *atomic force microscope (AFM)*, the scanning tip is pushed right up against the surface of the sample. When it scans, it moves up and down as it runs into the microscopic hills and valleys formed by the atoms present at the sample's surface. A variety of other scanning probe microscopes have been designed to detect properties such as friction, magnetic force, electrostatic force, van der Waals forces, heat, and sound.

One of the most important potential applications of scanning probe microscopy is the measurement of dynamic changes in the conformation of functioning biomolecules. Consider, for instance, how exciting it would be to "watch" a single enzyme molecule change its shape as it hydrolyzes ATP to provide the energy needed to transport ions across membranes. Such "molecular eavesdropping" is now entirely within the realm of possibility.

X-RAY DIFFRACTION ALLOWS THE THREE-DIMENSIONAL STRUCTURE OF MACROMOLECULES TO BE DETERMINED

Though **X-ray diffraction** does not involve microscopy, it is such an important method for investigating the three-dimensional structure of individual molecules that we include it here. This method reconstructs images from the diffraction patterns of X rays passing through a crystalline or fibrous specimen, thereby revealing molecular structure at the atomic level of resolution.

A good way to understand X-ray diffraction is to draw an analogy with visible light. As discussed earlier, light has certain properties that are best described as wavelike. Whenever wave phenomena occur in nature, interaction between waves can occur. If waves from two sources come into phase with one another, their total energy is additive (*constructive interference*); if they are out of phase, their energy is reduced (*destructive interference*). This effect can be seen when light passes through two pinholes in a piece of opaque material and then falls onto a white surface. Interference patterns result, with dark regions where light waves are out of phase and bright regions where they are in phase (Figure 39). If the wavelength of the light (λ) is known, one can measure the angle α between the original beam and the first diffraction peak and then calculate the distance d between the two holes with the formula

$$d = \frac{\lambda}{\sin \alpha} \qquad (6)$$

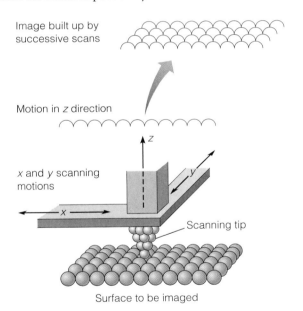

Figure 38 *Scanning Tunneling Electron Microscopy.* The scanning tunneling microscope (STM) uses electronic methods to move a metallic tip across the surface of a specimen. The tip is not drawn to scale in this illustration; the point of the tip is ideally composed of only one or a few atoms, shown here as balls. An electrical voltage is produced between the tip and the specimen surface. As the tip scans the specimen in the x and y directions, electron tunneling occurs at a rate dependent on the distance between the tip and the first layer of atoms in the surface. The instrument is designed to move the tip in the z direction to maintain a constant current flow. The movement is therefore a function of the tunneling current and is presented on a video screen. Successive scans then build up an image of the surface at atomic resolution.

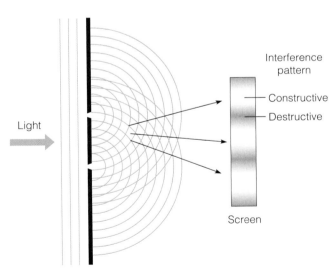

Figure 39 *Understanding Diffraction Patterns.* Any energy in the form of waves will produce interference patterns if the waves from two or more sources are superimposed in space. One of the simplest patterns can be seen when monochromatic light passes through two neighboring pinholes and is allowed to fall on a screen. When the light passes through the two pinholes, the holes act as light sources, with waves radiating from each and falling on a white surface. Where the waves are in the same phase a bright area appears (constructive interference), but where the waves are out of phase they cancel each other out, producing dark areas (destructive interference).

The same approach can be used to calculate the distance between atoms in crystals or fibers of proteins and nucleic acids. Instead of a sheet of paper with two holes in it, imagine that we have multiple layers of atoms organized in a crystal or fiber. And instead of visible light, which has much too long a wavelength to interact with atoms, we will use a narrow beam of X rays with wavelengths in the range of interatomic distances. As the X rays pass through the specimen, they reflect off planes of atoms, and the reflected beams come into constructive and destructive interference. The reflected beams then fall onto photographic plates behind the specimen, generating distinctive diffraction patterns. These patterns are then analyzed mathematically to deduce the three-dimensional structure of the original molecule. Figure 40 illustrates the use of this procedure to deduce the double-helical structure of DNA.

The technique of X-ray diffraction was developed in 1912 by Sir William Bragg, who used it to establish the structures of relatively simple mineral crystals. Forty years later, Max Perutz and John Kendrew found ways to apply X-ray diffraction to crystals of hemoglobin and myoglobin, providing our first view of the intricacies of protein structure. Since then, many proteins and other biological molecules have been crystallized and analyzed by X-ray diffraction. Although membrane proteins are much more difficult to crystallize than the proteins typically analyzed by X-ray diffraction, Hartmut Michel and Johann Deisenhofer overcame this obstacle in 1985 by crystallizing the proteins of a bacterial photosynthetic reaction center. They then went on to describe the molecular organization of the reaction center at a resolution of 0.3 nm, an accomplishment that earned them a Nobel Prize.

Figure 40 *X-Ray Diffraction.* X-ray diffraction can be used to analyze molecular structure at near-atomic resolution. ① X rays are diffracted by atoms in crystals or fibers just as light waves are diffracted by pinholes. In most cases, the specimens of interest to biologists are proteins or nucleic acids. The specific example illustrated in this figure is a DNA fiber. ② The resulting diffraction patterns are recorded photographically and can then be analyzed mathematically to deduce the molecular structure. This photograph depicts the actual X-ray diffraction pattern used by James Watson and Francis Crick to deduce the molecular structure of double-stranded DNA. ③ A computer graphic model of the DNA double helix.

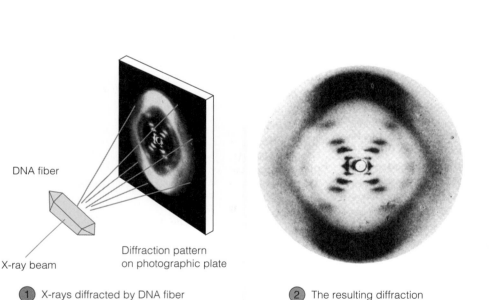

① X-rays diffracted by DNA fiber produce a diffraction pattern on a photographic plate

② The resulting diffraction pattern is analyzed mathematically

③ The three-dimensional structure of the molecule is deduced

Key Terms for Self-Testing

light microscope (p. 1)
electron microscope (p. 1)

Optical Principles of Microscopy
wavelength (p. 2)
interference (p. 3)
diffraction (p. 3)

focal length (p. 3)
angular aperture (p. 3)
resolution (p. 3)
refractive index (p. 4)
numerical aperture (NA) (p.4)
limit of resolution (p. 5)
useful magnification (p. 5)

The Light Microscope
compound microscope (p. 5)
condenser lens (p. 5)
stage (p. 5)
objective lens (p. 5)
ocular lens (p. 5)
intermediate lens (p. 5)

brightfield microscopy (p. 6)
phase-contrast microscopy (p. 6)
differential interference contrast (DIC) microscopy (p. 7)
fluorescence microscopy (p. 8)
fluorescence (p. 8)
immunofluorescence microscopy (p. 9)
indirect immunofluorescence (p. 9)
confocal scanning microscope (p. 11)
multiphoton excitation microscopy (p. 13)
digital deconvolution microscopy (p. 13)
digital video microscopy (p. 14)

Sample Preparation Techniques for Light Microscopy
fixatives (p. 16)
perfusion (p. 16)
microtome (p. 16)
staining (p. 16)
microscopic autoradiography (p. 16)

The Electron Microscope
transmission electron microscope (TEM) (p. 18)
electron gun (p. 18)
accelerating voltage (p. 18)
condenser lens (p. 18)
objective lens (p. 18)
intermediate lens (p. 18)
projector lens (p. 18)
electron micrograph (p. 19)
high-voltage electron microscope (HVEM) (p. 19)
scanning electron microscope (SEM) (p. 19)

Sample Preparation Techniques for Electron Microscopy
ultramicrotome (p. 21)
immunoelectron microscopy (p. 22)
negative staining (p. 22)

shadowing (p. 22)
vacuum evaporator (p. 22)
freeze fracturing (p. 23)
cryoprotection (p. 24)
intramembranous particle (IMP) (p. 24)
P face (p. 24)
E face (p. 24)
freeze etching (p. 25)
deep etching (p. 25)
stereo electron microscopy (p. 25)
critical point dryer (p. 26)
sputter coating (p. 26)

Other Imaging Methods
scanning probe microscope (p. 26)
X-ray diffraction (p. 27)

Suggested Reading

References of historical importance are marked with a •.

General References
Bradbury, S. *An Introduction to the Optical Microscope.* New York: Oxford University Press, 1984.
Slayter, E. M., and H. S. Slayter. *Light and Electron Microscopy.* New York: Cambridge University Press, 1992.

Light Microscopy
Allen, R. D. New observations on cell architecture and dynamics by video-enhanced contrast optical microscopy. *Annu. Rev. Biophys. Chem.* 14 (1985): 265.
Denk, Winfried, and K. Svoboda. Photon upmanship: Why multiphoton imaging is more than a gimmick. *Neuron* (1997) 18: 351.
• Ford, B. J. The earliest views. *Sci. Amer.* 278 (April 1998): 50.
Haugland, R. P. *Molecular Probes Handbook of Fluorescent Probes and Research Chemicals,* 5th ed. Eugene, OR: Molecular Probes, 1992.
Inoué, S., and K. R. Spring. *Video Microscopy: The Fundamentals,* 2d ed. New York: Plenum, 1997.
Lichtman, J. W. Confocal microscopy. *Sci. Amer.* 271 (August 1994): 40.
Mason, W. T., ed. *Fluorescent and Luminescent Probes for Biological Activity.* San Diego: Academic Press, 1993.
Potter, S.M. Vital imaging: Two photons are bettter than one. *Current Biology* (1996) 6: 1595.
Spencer, M. *Fundamentals of Light Microscopy.* Cambridge, England: Cambridge University Press, 1982.
Thomas, C.F., and J.G White. Four-dimensional imaging: The exploration of space and time. *Trends Biotechnol.* (1998) 16: 175
Wang, Y.L. Digital deconvolution of fluorescence images for biologists. *Methods Cell Biol.* (1998) 56: 305.

Electron Microscopy
Bozzola, J. J., and L. D. Russell. *Electron Microscopy: Principles and Techniques for Biologists.* Boston: Jones and Bartlett, 1992.
Flegler, S. L., J. W. Heckman, and K. L. Klomparens. *Scanning and Transmission Microscopy: An Introduction.* New York: Plenum, 1993.
Maunsbach, A. B., and B. A. Afzelius. *Biomedical Electron Microscopy: Illustrated Methods and Interpretations.* San Diego, CA: Academic Press, 1999.
Olson, A. J., and D. S. Goodsell. Visualizing biological molecules. *Sci. Amer.* 267 (November 1992): 76.

• Palade, G. E. Albert Claude and the beginning of biological electron microscopy. *J. Cell. Biol.* 50 (1971): 5D.
• Pease, D. C., and K. R. Porter. Electron microscopy and ultramicrotomy. *J. Cell Biol.* 91 (1981): 287s.
• Rasmussen, N. *Picture Control: The Electron Microscope and the Transformation of Biology in America, 1940–1960.* Stanford, CA: Stanford University Press, 1997.
• Satir, P. Keith R. Porter and the first electron micrograph of a cell. *Trends Cell Biol.* 7 (1997): 330.
Sommerville, J., and U. Scheer, eds. *Electron Microscopy in Molecular Biology: A Practical Approach.* Washington, DC: IRL Press, 1987.

Sample Preparation Techniques
Heuser, J. Quick-freeze, deep-etch preparation of samples for 3-D electron microscopy. *Trends Biochem. Sci.* 6 (1981): 64.
• Orci, L., and A. Perrelet. *Freeze-Etch Histology: A Comparison Between Thin Sections and Freeze-Etch Replicas.* New York: Springer-Verlag, 1975.
• Pinto da Silva, P., and D. Branton. Membrane splitting in freeze-etching. *J. Cell Biol.* 45 (1970): 598.

Other Imaging Methods
Engel, A., Y. Lyubchenko, and D. Müller. Atomic force microscopy: A powerful tool to observe biomolecules at work. *Trends Cell Biol.* 9 (1999) 77.
Engel, A., and D.J. Muller. Observing single biomolecules at work with the atomic force microscope. *Nat. Struct. Biol.* (2000) 7: 715.
Glusker, J. P., and K. N. Trueblood. *Crystal Structure Analysis: A Primer.* Oxford, England: Oxford University Press, 1985.
• Kendrew, J. C. The three-dimensional structure of a protein molecule. *Sci. Amer.* 205 (December 1961): 96.
Marti, O., and M. Amrein. *STM and SFM in Biology.* San Diego: Academic Press, 1993.
• Perutz, M. F. The hemoglobin molecule. *Sci. Amer.* 211 (November 1964): 64.
Roberts, C. J., P. M. Williams, M. C. Davies, D. E. Jackson, and S. J. B. Kendler. Atomic force microscopy and scanning tunneling microscopy: Refining techniques for studying biomolecules. *Trends Biotechnol.* 12 (1994): 127.
Stoffler, D., M.O. Steinmetz, and U. Aebi. Imaging biological matter across dimensions: From cells to molecules and atoms. *FASEB J.* (1999) 13 Suppl. 2: S195.

Index

A
Accelerating voltage, **18**
Analyzer, 7, 8f
Angular aperture of lens, **3**, 4f
Antibodies
    fluorescent, 9, 10
    gold-labeled, 22f
Antigens, 9
Artifacts, 16
Atomic force microscope (AFM), 27

B
Barrier filter, 9, 10f
Brightfield microscopy, **6**, 7t

C
Cathode, 18
Cold trap, 18
Compound light microscope, **5**, 6f
Condenser lenses
    electron microscope, **18**, 19f
    light microscopes, **5**, 6f
Confocal scanning microscopy, 7t, **11–13**
    laser, 12, 14f
Critical point dryer, **26**
Cryoprotection, **24**

D
Deep etching, **25**
Differential interference contrast (DIC) microscopy, **7**, 10f
    compared to other types of microscopy, 7t
Diffraction, **3**
Diffraction-limited spot, 4
Diffraction patterns, 27f
Digital deconvolution microscopy, 13, 15f
Digital video microscopy, **13–14**, 16f
Direct immunofluorescence, 9, 10f

E
E face, **24**, 25f
Electron beam, 1–2
Electron density, 19
Electron gun, **18**, 19f
Electron micrograph, 19
    radioisotopes and antibodies for localizing molecules in, 22
Electron microscope, **1**, 17–20
    optical system of, 2f
    resolution limits of, 4–5
    scanning, 19–20
    specimen preparation for, 20–26
    transmission, 18–19
Electrons, secondary, 20
Exciter filter, 9, 10f
Eyepiece of microscope, 5, 6f

F
Fixation, 16, 26
Fixatives, 16
Fluorescein, 9
Fluorescence, **8–9**
    principles of, 9f
Fluorescence microscopy, 7t, **8–11**
    fluorescent antibodies, 9–10
    fluorescent probes, additional, 10, 11f
    microscope, 9, 10f
    nature of fluorescence, 8, 9f
    traditional versus confocal, 12f
Fluorescent antibodies, **9–10**
Fluorescent probes, **9–11**
Focal length, 3
Four-dimensional microscopy, 14
Freeze etching, **25**
Freeze fracturing, **23–25**
    of plasma membrane, 24, 25f
    sectioning versus, 24, 25f

G
Green Fluorescent Protein (GFP), 10, 11f

H
High-voltage electron microscope (HVEM), **19**, 20f
Hoffman modulation contrast, 7
Hooke, Robert, 1

I
Illuminating wavelength of light, object size and visibility and, 1–3
Immersion oil, 5
Immunoelectron microscopy, **22**
Immunofluorescence microscopy, **9**, 10f
Indirect immunofluorescence, **9**, 10f
Interference, **3**
Intermediate lens
    electron microscope, **19**
    light microscope, **5**, 6f
Intramembranous particles (IMPs), **24**

J
Jassen, H., 1
Jassen, Z., 1

L
Laser scanning confocal microscope, 13, 15f
Leeuwenhoek, Antonie van, 1, 5
Lens
    condenser, and objective, 18, 19f
    electromagnetic, 18–19
    focal length, angular aperture, and resolution of, 3, 4f
    intermediate and projector, 18
    of light microscope, 5, 6f
    path of light through single, 13f
    useful magnification of, 5
Light, visible, 1
    path of, through single lens, 13f
Light microscope, **1**, 5–14
    confocal, 11–13
    differential interference contrast, 7, 8f, 9f
    digital video, 13–14
    fluorescence and immunofluorescence, 8–11
    optical system of, 2f, 3, 4, 5
    phase-contrast, 6–7, 8f
    preparation techniques for, 15–17
    resolution limits, 4–5
    types of, compared, 7t
Limit of resolution, 4, **5**

M
Microscopic autoradiography, **16**, 17f, 18f
Microscopy, optical principles of, 1–5
Microtome, **16**, 17f
Multiphoton excitation microscopy, **13**

N
Negative staining, **22**
Numerical aperture, **4**

O
Objective lenses
    electron microscope, **18**, 19f
    light microscope, **5**, 6f
Ocular lenses, **5**, 6f
Optical principles, **1–5**
    of differential interference contrast microscope, 8f
    illuminating wavelength related to size and visibility of objects, 1–3
    of light and electron microscopes, 2f
    of phase-contrast microscope, 8f
    resolution and discrimination of objects, 3–4
    resolution limits, 4–5

P
Perfusion, **16**
P face, **24**, 25f
Phalloidin, 10
Phase-contrast microscopy, **6**, 7t, 8f
Phase plate, 6, 8f
Plasma membrane, freeze fracturing of, 24, 25f
Polarizer, 7, 8f
Projector lens, electron microscope, **18**
Proteins, visualizing, with Green Fluorescent Protein, 11f

R
Radioisotopes, 22
Refractive index, **4**
Resolution of microscopes, **3–4**
    limits of, 4–5
Rhodamine, 9

S
Scanning electron microscopy (SEM), **19**, 21f
    specimen preparation for, 26
Scanning probe microscopy, **26–27**
Scanning tunneling microscope (STM), 26, 27f
Scintillator, 20
Secondary electrons, 20
Sectioning, 16
    with microtome, 17f
    ultrathin, 20–22
Shadowing, **22**, 23f
Source of illumination, 1
Specimen, 1
Specimen preparation for electron microscopy, 20–26
    fixation in, 26
    freeze fracturing and freeze etching, 23–25
    negative staining, 22
    radioisotopes and antibodies, 22
    sectioning vs. fracturing, 25f
    shadowing techniques, 22, 23f
    stereo electron, 25–26
    ultrathin sectioning and staining, 20–22
Specimen preparation for light microscopy, 16–17
    fixation, sectioning, and staining in, 16, 26
    microscopic autoradiography, 16–17, 18f
Sputter coating, **26**
Stage of microscope, **5**, 6f
Staining, 16
    negative, 22
    of ultrathin sections, 20
Stereo electron microscopy, **25–26**

T
Transmission electron microscope (TEM), **18**, 19f
    scanning electron micrographs versus from, 20f

U
Ultramicrotome, **21**
Ultrathin sectioning, 20–22
Useful magnification, **5**

V
Vacuum evaporator, **22**, 23f
Vacuum system of transmission electron microscope, 18
Viewing screen, electron microscope, 18
Visible light, 1
Voltage, electron microscope and, 19

W
Wave form, 2, 3f
Wavelength, **2**, 3f
Wollaston prisms, 7, 8f

X
X-ray diffraction, **27**, 28f